AF633468

Study Guide Crossword Puzzles

For English Literature

David J. Ragsdale, Ph.D.

Kingwood College

2003

Copley Custom Textbooks

An imprint of XanEdu Custom Publishing

Printed in the United States of America

ISBN 1-59399-026-X

Copley Custom Textbooks
An imprint of XanEdu Custom Publishing
300 North Zeeb Road
Ann Arbor, MI 48106
800-218-5971

Please visit our Web site at www.xanedu.com/copley

Contents

Geoffrey Chaucer

Chaucer Puzzle Clues

Across

2.) Parish clerk who receives an ignominious kiss.

6.) Western England resort city with mineral springs, this city also produces a flamboyant woman with a hat as big as a shield.

7.) Terry Jones's book suggest that the knight is actually one of these soldiers for hire.

12.) Carpenter John waits for the cry of this word to cut the ropes on his wooden tub hung high in the rafters.

15.) The rich color of the Wife's hose and the Summoner's bad complexion.

16.) The Wife of Bath had this number of husbands.

17.) The Squire's garment was embroidered with these.

20.) The number of archbishops that England had.

23.) A person (like Thomas Becket) who is killed for a movement or a cause is called a _____.

24.) This ancient Roman author said that men ought to marry women their own age.

25.) Chaucer's long narrative poem about a love relationship during the Trojan War was _____ and Criseide (probably based on Boccaccio). Shakespeare also wrote a play about this story.

28.) What saint said that proper monks should pore over books in their studies or work with their hands?

29.) A common personified abstraction in medieval times, this is the symbolic woman figure whose wheel represents the mutability of life's events.

31.) The number of stories that each pilgrim was originally supposed to tell according to the General Prologue (if Chaucer had completed the *Cant. Tales*).

32.) Medieval word meaning "creature" or "man."

34.) Sociologically and politically, the highest ranking pilgrim.

37.) To pay back or reciprocate.

38.) Medieval word for “named” or “called” in Chaucer.

39.) Medieval word meaning “also” in Chaucer.

40.) First name of the Wife of Bath’s last husband according to her Prologue.

Down

1.) A bawdy French story involving a practical joke; "The Miller's Tale" and "The Reeve's Tale" are both examples of this.

3.) How many sonnets did Chaucer write?

4.) His coat was threadbare, and his horse was thin, demonstrating that earthly luxuries and prestige did not mean much to him.

5.) The name for any man whose wife has an affair behind his back. According to legend, medieval men who became one of these sprouted little horns and could hear a cuckoo bird singing in the distance.

8.) The sole, powerful religion in medieval England.

9.) With Chaucer's squire as with Boccaccio, this bird sometimes symbolizes lovers because it is active at night.

10.) The last name of the archbishop killed in Canterbury Cathedral about 200 years before Chaucer.

11.) The one pilgrim clergyman whose life is pure and righteous.

13.) Because he had been a carpenter in his youth, this pilgrim is offended by the Miller's story about a gullible carpenter and tells a bawdy one himself (though not in our text) to get back at the Miller. Conflicts like this one are the "links" that give some cohesion to the tales.

14.) The relationship of Alisoun to Carpenter John.

16.) The century in which Chaucer lived and wrote.

18.) Is Canterbury east or west of London?

19.) The corrupt clergyman pilgrim whose job was to notify and bring canon law offenders to an ecclesiastical court. (Use the Middle English spelling of his name.)

21.) The Miller frequently won a ram for winning at this activity.

22.) Crude, boisterous pilgrim with large nostrils and a nose wart.

26.) Corrupt pilgrim guilty of simony for taking money in exchange for bogus church forgiveness.

27.) Born in Greece but resident in Egypt, this second-century astronomer gave the medieval world its model of the earth as the center of the universe. A model of his universe diagram is located in the back of the *Norton Anthology*.

28.) The month in which the pilgrims leave on their trip to Canterbury.

29.) A _____ tale or story involving a story within a story, such as *The Wizard of Oz,* Coleridge's "Rime of the Ancient Mariner," or Twain's "The Celebrated Jumping Frog of Calaveras County."

30.) Name of the Southwerk inn where the pilgrims assembled.

33.) The son of the knight.

35.) According to line 24, there were twenty- _____ pilgrims in the company/group.

36.) The adjective which Chaucer uses in "The Miller's Tale" to tell us that Nicholas is handy, opportunistic, and close at hand.

Sir Gawain and Medieval Drama

Sir Gawain and Medieval Drama Puzzle Clues

Across

5.) Bercilak de Hautdesert lives at the Green _____.

6.) Medieval *miracle* plays focus on the lives of _________.

7.) According to a legend (alluded to at the beginning and end of *Sir Gawain*), this Roman founded Britain.

12.) Home of Arthur and his chivalric knights.

14.) One of the leading towns for medieval mystery play pageants.

16.) A hunted animal symbolizing coy, demure behavior.

17.) Mystery plays typically occurred on these moveable stages.

18.) Was the *Sir Gawain* poem more a test of physical strength or of truth?

20.) The 5-line rhyme scheme for the bob and wheel.

21.) This feature involving repeating consonants was a cohesive device and also an embellishment for early poetry versification before modern rhyme and meter.

23.) This character with an account book introduces his knowledgeable sister to Everyman.

25.) To pay back either in love or in war.

26.) Character symbolizing self-awareness who says, "Everyman, I will go with thee and be thy guide, in thy most need. . . ."

27.) What number enumerates both the number of Gawain's senses and the number of Christ's wounds on the cross?

28.) Gawain's meeting deadline is a year and a ______.

Down

1.) Before Gawain steps in, this virtuous person is the first person who is challenged by the Green Knight.

2.) Based on "The Second Shepherd's Play" and "Everyman," which of the 3 types of medieval drama contains more comedy?

3.) Was *Sir Gawain* written in the same century as *The Canterbury Tales*?

4.) The name of the special refrain structure portion in *Sir Gawain* that has 2 syllables.

5.) "Canon" law refers to the rules and statutes of the _____.

8.) *Everyman* is what type of medieval play?

9.) This person tells Everyman that he can't go along because he has a cramp in his toe.

10.) The chief of the deadly sins.

11.) What holiday is being celebrated during the time in which the Green Knight visits Arthur's kingdom?

13.) The sheep stealer who disguises a lamb as a baby.

15.) A five-sided star which in medieval times symbolized Solomon's wisdom.

17.) A key medieval drama town in which a Master wrote *The Second Shepherd's Play.*

19.) An extended metaphorical narrative in which a figure stands for a specific quality, such as Holinesse.

22.) The number of hunts for Sir Gawain's castle host.

24.) The number of deadly sins in medieval thinking.

27.) The hunted animal which symbolizes craftiness.

The Tudors

1
2
3
4
5
6
7
8
9
10
11
12
13
14
15
16
17
18
19
20
21
22
23
24
25
26
27
28
29
30
31
32
33
34
35
36
37
38
39
40
41
42
43
44
45

The Tudors Puzzle Clues

Across

4.) Sometimes used in her name, this is actually the region of Spain where Henry's first wife was from.

5.) The surname of Henry VIII's 6 th wife.

6.) The last name of the high ranking secular official/clergyman and author of *Utopia* who was beheaded for refusing to recognize Henry as the head of the English Church.

7.) The first name of the son who succeeded Henry VIII as king of England.

10.) The major river that runs through London; the Tower of London and the Parliament building are located on its banks. Consult the map of London located in the endpapers inside the back cover of your textbook.

12.) Last name of the Puritan firebrand who led the English parliament during 1649-1660 when England had no king.

14.) Catherine Howard was wife number _____.

17.) Rounding it off to the nearest decade, Henry VIII reigned for nearly _____ years? (e.g., twenty, thirty, forty, fifty, etc.)

18.) How many children did Elizabeth I have?

22.) For his fourth wife, Henry married Anne, a princess from the German dukedom of _____.

23.) The first name of Henry VIII's father was _____.

24.) First wife Catherine's parents were Isabella and Ferdinand of the country of _____.

25.) Sir Walter Raleigh, Anne Boleyn, Thomas More, and even a young Elizabeth herself were all at various times prisoners incarcerated in the ______ of London.

26.) Mary Tudor ultimately married _____ of Spain, but their marriage was childless and largely unhappy.

28.) The first name of the man who succeeded Elizabeth I in 1603 and started the Stuart dynasty. He was the son of Mary Queen of Scots.

30.) A minor poet and a soldier, this man was a favorite of Queen Elizabeth.

31.) The position of Catholic Clement VII, with whom Henry VIII severed all relations during the 1530's.

32.) Of all of Henry VIII's 3 children, the one who ruled the best and the longest.

36.) Henry VII started the Tudor dynasty in 1485, thereby ending the almost 50-year upheaval called The Wars of the _____.

38.) Many people who became religious martyrs during the religious persecution of Bloody Mary were chronicled in a Book of Martyrs written by John ______ . There's a short excerpt in our textbook.

40.) The largest city in England, home to Westminster Abbey, where English monarchs are crowned and buried, and St. Paul's Cathedral.

41.) Elizabeth finally and reluctantly gave the order to execute this annoying Catholic queen relative to the north who kept trying to overthrow her. Ironically, this Catholic queen's son eventually succeeded Elizabeth upon her own death in 1603.

42.) Henry remarried several times, not because he was more lecherous than other men, but because he wanted a wife who could give him a ____.

44.) The allies of the Tudors during the wars in 36 down.

45.) Queen Anne Boleyn and Queen Catherine Howard were both ultimately _____, but so were Thomas More, Thomas Cromwell, 2 Seymour men, and Cardinal Wolsey.

Down

1.) Anne Boleyn was beheaded on false adultery charges, but Henry did her the favor of having her beheaded with a _____ rather than by the standard axe, which was not always efficient, quick, and painless to do its work.

2.) One of the highlights of Elizabeth's reign was when she and Raleigh defeated the menacing Spanish ______ in 1588.

3.) Mary Tudor earned her nickname of Bloody Mary because she decided to behead or _____citizens who refused to take an oath of loyalty to Catholicism.

4.) Catherine, Henry VIII's first wife, was first married to _____, Henry's older brother who died at 16, only 6 months into the marriage.

8.) One of the most powerful men during Henry VIII's reign was Cardinal _____ although he was eventually beheaded too.

9.) Catherine of Aragon and Anne of Cleves were both _____ from Henry rather than being executed.

11.) The daughter of Catherine of Aragon and Henry VIII.

12.) Jane Seymour and Catherine Parr both died in [the condition of] _____; Jane did so during Henry's lifetime, Catherine after it.

13.) Anne Boleyn was queen of England for only about 36 _____.

14.) The Catholic country across the English Channel from England. Henry VIII made some attempts at being friendly with this country, such as the Field of the Cloth of Gold, but this country tried to unseat Elizabeth several times during her reign decades later.

15.) After almost a decade of passing statutes reducing the power of the Catholic Church in England, in 1539 Henry VIII severed the last tie with Rome when he closed all the __________ in England.

16.) The last name shared by Henry VIII's 5th wife as well as the poet, the Earl of Surrey.

17.) Henry VIII alleged that Anne Boleyn had adulterous affairs with several men including an incestuous relationship with her brother, but these charges are probably ______. Nevertheless, they were all executed.

19.) Shortly before he died, Henry VIII gave orders that he was to be buried next to Jane Seymour, his wife number _____.

20.) Because the pope would not grant him a divorce from his first wife, Henry VIII passed a series of statutes in the decade of the 1530's so that England would no longer be a _____ country with sovereignty to Rome.

21.) After closing the monasteries in 1539, Henry VIII dictated that, in terms of religion, England would now be _____.

27.) In 1515, Martin Luther started the ______ Reformation by challenging the doctrine and practices of the Catholic Church by nailing a list of Church abuses to the door of the church in Wittenburg, Germany.

29.) 6 across was executed because he failed to sign this statute, The Act of _____, in which Henry VIII said that he was now becoming the head of the Church of England.

33.) A century after Elizabeth died, rival France was ruled from 1715-74 by Louis the ______.

34.) How many wives did Henry VIII have?

35.) Henry VIII and his daughter Elizabeth were both concerned at times that someone would try to assassinate them by secretively getting them to ingest this.

37.) The last name of the wife who finally gave Henry VIII a son.

39.) Henry VIII was married to one of his wives for 20 years, but she was then displaced by a young, aggressive, dark-haired woman with the last name of _____.

43.) In the period of instability known as the Wars of the Roses, the ______ (red or white?) was the rose that represented the side shared by the Lancasters and Tudors. This tumultuous time was settled when Henry VII established the Tudor dynasty and succession in 1485.

Edmund Spenser

Edmund Spenser Puzzle Clues

Across

2.) A literary work in which a literal level parallels and has a 1-to-1 correspondence with another deeper, symbolic one which may be religious or political.

7.) The first part of the Greek god Apollo's name; he is responsible for the beautiful, "rosy-fingered dawn" expression which epic poems celebrate in their typical dignified style.

8.) A descent to this is one of the traditional characteristics of an epic poem.

10.) The nonChristian saracen Sans Foy lives without this, as his name tells readers.

11.) This animal accompanies and symbolizes Lechery in Book IV's House of Pride.

12.) In canto III of *The Faerie Queene,* Una encounters this negative female character, whose daughter Abessa is having an amorous relationship with 13 down.

14.) Dressed meretriciously in scarlet, Duessa exhibits echoes of the Whore of Babylon, who appears in the book of _____, which is the last book in the Bible.

16.) This classical Roman writer of the epic *Aeneid* was also one of the first people to write pastoral poems and also agricultural poems called georgics.

17.) An epic poet traditionally invokes the assistance of a _____, one of nine female deities who oversee the arts.

18.) The number of *Faerie Queene* books that Spenser finished before he died.

20.) Former lover Fradubio was changed into this when he saw Duessa bathing in her original hag-like form.

21.) Another word for greed, one of the seven deadly sins.

22.) The seemingly sober color of Archimago's robes when Una and Redcrosse Knight first meet him.

24.) In keeping with the epic tradition of catalogues, the FQ includes a lengthy list of _______ early in book I, canto I.

25.) The color of Una, her donkey, and her lamb, signifying her spiritual purity.

27.) The primary male villain of FQ book I.

29.) The hero who symbolizes Holinesse.

31.) An archaic name for clothing, especially clothing for widows or for shepherds. (Consult the 1st page of the poem.)

32.) Meaning changeableness, this is the subject of 2 fragmentary FQ cantos which suggest that everything is subject to change.

34.) Did Spenser live to a later date than Elizabeth I?

35.) Britomart, the heroine of FQ book III, symbolized this virtue according to Spenser's letter to Raleigh.

37.) This animal, which accompanies and protects Una for awhile in canto III, can sometimes symbolize royalty and at other times symbolize anger, as in Canto IV.

38.) She is half woman and half serpent and lives in a dark cave with her misshapen offspring. (Use the British spelling.)

39.) This unChristian Saracen lives without law, as his name tell us.

Down

1.) This traditional pastoral name is one of the shepherd characters in "The Shepherde's Calender" [sic].

3.) A shepherd traditionally plays music on _____ pipes, which Spenser says he is abandoning at the beginning of *The Faerie Queene*.

4.) Spenser writes a letter to this gentleman (and a favorite of Elizabeth I) explaining much of his complex intent in the *FQ*.

5.) This member of the 7 deadly sins eats more food than he can possibly use.

6.) 9 down holds one of these in her hand in canto 4 so she can proudly reflect on her favorite subject: herself

7.) The chief of the deadly sins, an idea which goes all the way back to the Garden of Eden, is often depicted in literature as causing a person to be inflated with his own importance. What is the chief deadly sin?

8.) Redcrosse Knight's pure companion who rides on a humble donkey.

9.) She rules the House of Pride and is accompanied by 6 vice-like advisors.

13.) His name means "church raper," and he lives with 12 across until Una's lion kills him in canto III.

14.) This color, which is the hue of the Roman god (and planet) Mars as well as the tawdry promiscuity of Duessa, stands in stark contrast to Una's color of purity.

15.) Duessa ironically goes by this virtuous sounding name when she accompanies Sans Foy and, even earlier, Fradubio.

16.) This nauseating, emetic product of 38 down contains books and pamphlets of Catholic propaganda and dogma, thus solidifying the FQ's religious satire

18.) The rural vocation of the speaker in a pastoral poem.

19.) First established by Virgil and others, this is a type of poem that celebrates simple, pure, country life taking care of sheep on the hillsides; often uses stock, archaic names.

23.) Una has one of these animals that she takes with her; it also symbolizes sacrificial purity and Christianity at times.

26.) Another name for pastoral poems.

28.) A villainess in *The Faerie Queene*, she stands for duality and two-faced behavior, as her name indicates.

30.) One of the nonChristian saracens in FQ, this evil knight embodies living without faith.

32.) In canto IV's House of Pride, the large edifice itself is built cunningly with bricks though the bricks are not actually secured together with this.

33.) When Redcrosse Knight breaks a tree limb off to make Fidessa a garland, the tree does this.

36.) How many sprites does 27 across magically conjure up to help him confuse and disrupt Redcrosse Knight and Una?

William Shakespeare

Shakespeare Puzzle Clues

Across

1.) The name of the English monarch who was ruling during the first 39 years of Shakespeare's life.

4.) In Act I, scene I, Lear cautions Cordelia to mend her speech because " ______ comes of nothing."

6.) In a famous *Julius Caesar* oration, Marc Antony says, "Friends, Romans, countrymen, lend me your _____" [singular].

10.) In *Hamlet*, Polonius gives his son the advice, "To thine own self be _____."

13.) The name of the major tragedy in which a Moorish general becomes terribly jealous.

14.) One of Lear's two wicked daughters.

15.) Lear laments toward the middle of the play that he "is a man more sinned against than _____."

17.) This unfortunate King Lear character is a father of two who has his eyes put out.

19.) Shakespeare wrote 154 of these.

20.) A play like "Measure for Measure," which is somewhat comic but also disturbingly painful at other times is called a bitter comedy or a _____ play.

22.) Shakespeare's mother's maiden name; also the name of a forest in *As You Like It.*

23.) In an Act III soliloquy, Lear tells pomp to take this.

24.) He collaborated with Heminges to publish the First Folio of Shakespeare's work in 1623.

26.) Prince Hal says in *Henry IV, Part I* that if every day were a holiday, then to _____ would be as tedious as to work.

28.) The bard's wife's maiden name.

31.) A disyllabic poetic foot with its first syllable unaccented and its second syllable accented.

33.) Unrhymed iambic pentameter.

35.) Hamlet says derisively to his mother that "______, thy name is woman."

36.) To disguise himself, Edgar feigns insanity and masquerades as a rural beggar named Poor _____.

37.) According to the 3 dramatic unities, specifically the unity of time, all the events in a play should take place within the time span of one _____.

39.) How many lines are there at the end of every Shakespearean sonnet that rhyme only with each other with the rhyme scheme "g"?

40.) When Regan says that she dearly loves her father and when Edmund says that brother Edgar hates his, both evil characters are telling _____.

41.) Hamlet puns in Act I, scene ii that King Claudius is "a little more than _____ but less than kind."

42.) The foolishly tragic king who retires but gives his kingdom away to his two wicked daughters and their wicked husbands.

Down

2.) Shakespeare's canon consists of 3 types of plays: tragedies, comedies, and _____ (also called chronicle plays).

3.) This Roman tragedian probably originated the ideas of a revenge tragedy and of an avenging ghost that serve as models for Shakespeare. Remember that Shakespeare could read the Roman playwrights in Latin; however, he could not read Greeks dramatists like Sophocles because he could not read Greek since he never went to college.

4.) In Julius Caesar, Marc Antony tells a crowd that he "comes to bury Caesar, ____ to praise him."

5.) She marries the man who killed her royal husband and doesn't seem to sense a problem in her first husband's demise.

7.) Hamlet wants to seek _____ against Claudius for killing Hamlet's father.

8.) Lear's _____ is his ironic and comic companion who says things to Lear that no one else can get away with. His last words as he disappears in the middle of the play are "And I'll go to bed at noon."

9.) This Elizabethan playwright, second only to Shakespeare in dramatic ability, wrote *Every Man in his Humour* and *Volpone.*

11.) Every Shakespearean tragic hero is of _____ social status.

12.) When the character in 17 across is blinded, he is cruelly told to ____ his way to Dover.

16.) Poor people who paid the smallest amount to get into an Elizabethan theater just stood on the ground for the whole play and were, therefore, called _______.

17.) The name of the famous theater in which Shakespeare's plays were performed and in which he was a partial owner.

18.) The name of Shakespeare's last play, which is set on an island on which people (especially 2 brothers) are stranded after a terrible storm.

20.) Poems consist completely of verse, but _____ often have both poetry and prose in them.

21.) Lear warns Cordelia to "mend [her] speech lest it _____ her fortune."

25.) One of Shakespeare's major tragedies, this play involves a Scottish nobleman who murders his way to the throne but is warned to beware when Birnam Wood comes to Dunsinane.

27.) Shakespeare wrote ______-seven plays.

29.) In Act III, King Lear is in anguish when he cries, "Blow _____, crack your cheeks."

30.) All Shakespearean plays have this number of acts in them.

32.) The name of the river that runs through London and right by the location of the theater in 17 down. Theaters were built across this river from London in order to be out of the Sheriff of London's autocratic jurisdiction.

34.) Edgar's wicked illegitimate brother who speaks a soliloquy toward nature about how unfair life is.

35.) Shakespeare's First _____, the first collected edition of his works, was published in 1623, which was 7 years after his death.

38.) The first four lines of an Italian sonnet have this rhyme scheme.

The Seventeenth Century (excluding Milton)

17th-Century (excluding Milton) Puzzle Clues

Across

3.) When a valediction speaker says that the love between his wife and him will not fall apart, he likens it to this metal which merely expands when beat "to airy thinness."

6.) When Elizabeth I died in 1603, the Tudor dynasty ended, and the _____ dynasty began.

7.) In Meditation 17, Donne coined the phrase "for whom the bell ______," which Hemingway later adapted as a title for his 20th-century novel.

9.) A Marvell speaker says that he doesn't have all the time in the world because, behind him, he can hear time's winged _______.

12.) _____ the 1st was the first monarch in the new dynasty mentioned in 6 across.

13.) Marvell's speaker says that the grave is "a fine and private place but none I think do there _____."

15.) The 16 across sonnet ends ironically by saying that, through Christ, death itself will ultimately _____.

16.) Donne's most famous sonnet begins "____ be not proud, though some have called thee/ Mighty and dreadful, for thou art not so."

17.) A famous Renaissance theme is _____ diem, which means to seize the day.

19.) "The Sun Rising" twists the Ptolemaic universe model of the sun at the universe's center and says that, for these 2 lovers, their _____ is the center of the sun's universe.

20.) One of Donne's well-known Holy Sonnets refers to the Christian concept of the Trinity as it begins, "Batter my heart, 3-personed _____."

21.) An aubade is a poem about the what time of day?

23.) John Bunyan wrote *The _____ Progress*, which our anthology (in its intro of Bunyan) describes as "the most popular allegory in our literature."

24.) According to the theme in 16 across, Herrick suggests that it would be wise for virgins to gather ______ while they may.

26.) The result of a canonization (see the Donne poem, "The Canonization") is that a person becomes one of these.

27.) This first English poet laureate wrote "MacFlecknoe" and died in 1700.

29.) This person, Oliver _____, spearheaded the Puritan-led Parliament which waged civil war against Charles I, eventually beheaded Charles I, and ruled England without a king from 1649-1660.

30.) This writer was probably the second most famous Elizabethan dramatist after Shakespeare and wrote such plays as *Volpone* and *Every Man in his Humour*.

Down

1.) The name for the period 1660-1700, which derived its name from the king being returned to the throne after a decade's absence. The period is famous for its witty, though sometimes bawdy, plays.

2.) In 1666 London was heavily damaged by _____.

4.) This English philosopher and author of "An Essay Concerning Human Understanding" was about the only English intellectual actually to be involved in The Enlightenment that was going on in Europe.

5.) Writer of shaped poems such as "Easter Wings" and "The Altar," this man wrote only devotional, religious poems.

8.) In "MacFlecknoe," this inferior writer is the satiric target who inherits the crown of dullness from Flecknoe.

10.) An epithalamium is an "occasional" poem written to commemorate the special occasion of a _____.

11.) Donne wittily uses this insect as a central illustration in a poem about a man trying futilely to get a woman to spend the night with him.

14.) "A seemingly contradictory statement that may nevertheless be true"—a figure of speech used several times by Donne and the other metaphysical poets.

18.) This man wrote "about 1.3 million words," including a famous, detailed diary of the 1660's which describes London's great conflagration.

19.) In 1610, King James I commissioned a new translation of the _____, a translation which became the leading translation of Protestants for 350 years.

21.) To do something impossible, Donne suggests that someone "get with child" this human-shaped plant root.

22.) To illustrate and reinforce the connection between all people, Donne used the metaphor in one of his prose works that "no man is an _____."

25.) Donne was accomplished at writing religious versions of this type of 14-line poem.

28.) Marvell's speaker tells his coy mistress that, even though they can't make time stand still, they can still make him _____.

John Milton

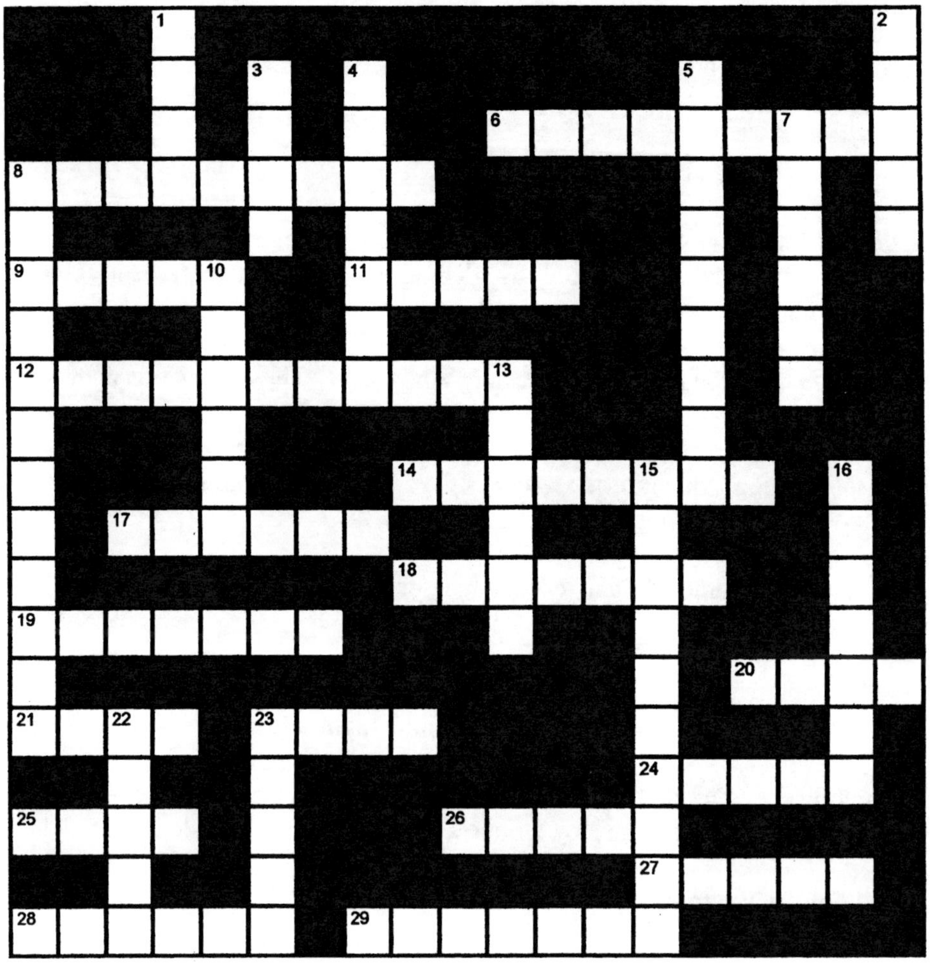

Milton Puzzle Clues

Across

6.) *In Paradise* Lost, Satan is described as being as large as a sea-beast or whale called this. This word is also a notable Renaissance work by Hobbes.

8.) In Hebrew, the word "Satan" means _______. (Consult the text's footnotes.)

9.) In *Paradise Lost (PL)*, after cataloguing the initial group of demons, the narrative goes on to catalogue a group including Osiris, Isis, and Orus, who are all from the country of _____.

11.) Milton was married _____ times, not because divorce was an option but because his previous wives died.

12.) The word that Milton coined himself to label the palace of the demons toward the end of *PL*, book I.

14.) The architect of 12 across and also sometimes another name for Vulcan.

17.) Any epic writer must do his research homework by reading *The Iliad* and *The Odyssey* by Homer, a writer from the country of _____.

18.) Milton's great pastoral elegy.

19.) Is the sonnet in 27 across English or Italian in its rhyme scheme?

20.) The sonnet in 27 across ends with the famous line that "They also serve who only stand and _____."

21.) What is the rhyme scheme for the first 4 lines of a Petrarchan sonnet?

23.) Every epic begins with an invocation to the _____.

24.) The glossary in the back of our textbook defines a ____ as "a formal, and usually long, poetic lament for someone who has died." Milton wrote his on the death of a college friend, Edward King.

25.) A long narrative poem that contains characteristics such as a descent into the underworld, supernatural intervention, a catalogue, battles, lofty diction, and a beginning in medias res.

26.) _____ Pendragon, mentioned in PL, book I, was the father of King Arthur.

27.) Probably Milton's most famous sonnet is "When I Consider How My Light is _____."

28.) An epic usually begins in _____ res, which means "in the middle of things."

29.) The biographical intro on Milton in our textbook points out that, between 1643 and 1645, he tried getting solace from his "disastrous marriage" by writing several works on the subject of _____ though these writings "were denounced in Parliament, from pulpits, and in print."

Down

1.) Satan and his army of demons are cast out of heaven and finally land in a lake of ____.

2.) Also called Mount Oreb, this Hebrew mountain was special to Moses, whom Milton addressed as the Judeo-Christian alternative to the pagan muse of early epic poetry.

3.) Satan's spear is equal in size to a tallest Norwegian pine tree that was ever cut down to form the _____ of a great admiral's ship.

4.) Milton was a _____, a strict, conservative Christian sect that wielded great political power during 1650-1658 and even brought some religious settlers to America.

5.) Another name for a list of things such as trees, warriors, rivers, etc.; one of the characteristics of an epic poem.

7.) The family estate where Milton was able to study and read for several years in preparation and training for his poetic career.

8.) The name of Milton's great prose defense of freedom of the press.

10.) The number of books in *Paradise Lost* or virtually any epic.

13.) The first demon follower of Satan and Beelzebub, described as a "horrid king besmeared with blood/ Of human sacrifice, and parents' tears" because children were sacrificed to him.

15.) PL is written in this, which is unrhymed iambic pentameter. Plays had used this verse form before, but Milton popularized it for regular poetry now.

16.) Milton promised as his thesis in PL to "_____ the ways of God to man," thereby making the poem a theodicy.

22.) Like Homer before him, Milton suffered from a severe physical disability when he went ____ in 1653, but he was able to write PL with the help of amanuenses such as his daughters.

23.) This great Hebrew writer wrote the first 5 books of the Bible. He became Milton's muse in a way because Milton wanted to follow the example of Homer but to Christianize the poem. Homer was not a Christian because he lived 800 years before Christ was born.

The Eighteenth Century

18th Century Puzzle Clues

Across

2.) He co-authored *The Spectator* with Addison.

3.) The surname of a man with the first name of Lemuel.

4.) John Gay wrote "The Beggar's _____."

5.) He co-wrote *The Spectator* from 1709-1711 and then returned to it in 1714 even though 2 across didn't.

7.) One of the leading painters of the century, he painted the well known "Blue Boy."

13.) The periodical essay which observed London social customs , provided gentle satire, and contained a club consisting of Sir R. de Coverley, Sir Andrew Freeport, and Will Honeycomb.

15.) De Coverley's first name.

17.) The "neoclassical" label for the 18th century means that the era tried to revive an interest in the arts, letters, and architecture from the much earlier _____ and Roman periods.

18.) He wrote an "Essay on Criticism."

20.) What is the rhyme scheme of the first 6 lines of any poem written in heroic couplets?

22.) What German-born composer who is buried in Westminster Abbey and wrote the oratorio "The Messiah" that has the famous "Hallelujah Chorus" in it?

23.) The controlling persona of *The Tatler* was the fictional character Isaac _____.

27.) Sir Robert ______ was British prime minister from 1721-42 while his son, Horace _______, wrote *The Castle of Otranto*, which was one of the very first gothic novels. Gothic novels first appeared late in the 18th century and tried to include elements of horror and supernaturalism into fiction, paving the way for the upcoming Romantic period. [Mary Shelley's *Frankenstein*, Bram Stoker's *Dracula*, and many Edgar Allan Poe short stories are later gothic tales as well.]

29.) Another name for the 18th century, particularly the early half of it. Derived from the Roman emperor Caesar Augustus.

31.) The periodical essay series written by Samuel Johnson between 1750 and 1752.

33.) The 18th-century novelist who wrote the comic, episodic novels *Joseph Andrews* and *Tom Jones.*

35.) Samuel Johnson's greatest poem is "The Vanity of Human _____," which analyzes people's tendency to fear the wrong things and wish for other wrong an deceptive goods and abilities that will never bring them true happiness.

36.) The 18th-century novelist who wrote the epistolary novel *Humphry Clinker.*

37.) Oliver ______, a second-tier writer who wrote *The Vicar of Wakefield, The Deserted Village*, and *The Citizen of the World* periodical essays. Samuel Johnson once said that if he were ever trapped under a shelter by a rainstorm, he'd rather have this man's company and conversation than anyone else's.

39.) *The Citizen of the* _____ essays in 37 across were written in mid-century but nevertheless followed in the periodical essay tradition largely begun by *The Spectator* in 1709.

40.) _______ includes a number of literary ways of holding up a subject and making it "laughable or contemptible" by diminishing it or exaggerating it. Irony is often a part of this technique, and the technique can be either gentle (Horatian) or bitter (Juvenalian).

Down

1.) This 18th-century artist often drew satirical prints such as "The Rake's Progress," "The Harlot's Progress," and "Beer Street" and "Gin Lane."

2.) This 18th-century novelist wrote *Tristram Shandy*.

3.) This poet later in the century wrote "Elegy Written in a Country Churchyard."

6.) This 18th-century novelist, though often not accepted by other novelists of the century, nevertheless wrote *Moll Flanders* and *Robinson Crusoe*.

8.) The young Scottish author of the Life of Johnson, one of the greatest biographies in English.

9.) Addison and Steele co-authored this essay periodical in the years after *The Spectator*. Its leading characters were Mrs. Aspasia Lizard and the Lizard family.

10.) He wrote MacFlecknoe, died in 1700, and was among the first poet laureates of England.

11.) She wrote *A Serious Proposal for the Ladies* and *Some Reflections on Marriage* (an excerpt is in our book) and was one of the first people to speak out and encourage women to get an education and do more with their lives.

12.) David _____ was once a student of Samuel Johnson, and the 2 of them walked to London together in 1737. This man later went on to become the manager of the Drury Lane Theatre in London, where Johnson's only play, *Irene,* was produced. *The Norton Anthology* calls him "the leading actor of his generation."

14.) Minor late-century poet who wrote "The Task."

16.) Having characteristics such as a descent into the underworld, a catalogue, and a beginning in medias res, this subgenre of poem was almost nonexistent in the 18th century.

19.) He dominated English literature during the second half of the century and was remarkably versatile, writing critical biographies, poetry, a novella, a play, and a dictionary.

21.) This peer ignored Johnson when the good doctor visited him and asked for patronage in 1748 but later offered sponsorship when Johnson's *Dictionary* was already finished, prompting Johnson to write a famously tactful but cold letter and likening the help to throwing a drowning man a life preserver after he's already crawled up on the beach.

24.) Almost every poem written in England in the first half of the 18th century is written in heroic _____.

25.) Lines of poetry containing 5 feet are called __________.

26.) In 1714, the Hanoverian dynasty (originating from Germany) began, and England had 3 kings in a row with the first name of ________.

28.) The monarch who died in 1714 was Queen _____, whose name is usually associated now with a type of elegant dining room furniture.

30.) He was the leading prose satirist of the century.

31.) This painter, Sir Joshua ______, was a member of Johnson's Club and one of the most famous artists of the century.

32.) This word can have 2 meanings. One meaning is the name of the man for whom Samuel Johnson did so much work on *The Gentleman's Magazine*; another meaning for the word is a synonym for a grotto (which Alexander Pope had at his home in Twickenham).

34.) The last name of Charles and his brother, who founded the Methodist denomination in 18th-century England and wrote sermons and songs.

38.) Pope wrote an important poem titled "Essay on ____," which dealt with the place of human beings in the cosmic order.

Alexander Pope

Alexander Pope Puzzle Clues

Across

1.) In a line that echoes Milton, Pope says that the purpose of his greatest poem is "to vindicate the ways of God to ___."
4.) "____ nature and ____ sense must ever join" is the line that precedes 10 across.
6.) the sacred mountain of poetry in Greece
7.) "Why has not man a microscopic eye?/ For this plain reason, man is not a _____."
10.) "To ___ is human, to forgive divine."
11.) "_____ springs eternal in the human breast:/Man never is, but always to be blest."
12.) One of the authors of *The Spectator* who got Pope to write rare essays for that essay periodical. He also approved of Pope's early "Essay on Criticism."
16.) "First follow _____ and your judgment frame/ By her just standard, which is still the same."
18.) "In ____, in reasoning ____, our error lies;
All quit their sphere, and rush into the skies.
_____ still is aiming at the blest abodes,
Men would be angels, angels would be gods."
20.) Pope alludes frequently to the Great Chain of _______.
21.) "The Rape of the Lock" is a ____ ______ in terms of genre.
23.) In 1728, Pope wrote this satire lampooning the poet Theobald as the king of dunces.
24.) "A little learning is a dangerous thing;
Drink deep, or taste not the _____ spring.
There shallow draughts intoxicate the brain,
And drinking largely sobers us again."
25.) The ______ Age--a name for the 18th-century, especially the first half of it.
26.) "True ____ is Nature to advantage dressed,
What oft was thought, but ne'er so well expressed;
Something whose truth convinced at sight we find,
That gives us back the image of our mind."
27.) "An Essay of Criticism" refers to this early writer and writing model as "the Stagirite."

Down

1.) In the third line of "Rape of the Lock," Pope addresses this being that any serious epic will address and invoke for inspiration as well.
2.) Pope said that any good critic should do his homework and know the early, important epic work of this writer: "Be _____'s works your study and delight,/Read them by day, and meditate by night."
3.) "All seems infected that the infected spy,
As all looks yellow to the ________ eye."
5.) Wispy, ethereal, supernatural beings capable of flight in "Rape of the Lock."
8.) "A little ______ is a dangerous thing."
9.) Pope wrote almost exclusively in heroic couplets, which are rhyming pairs of _____ pentameter.
11.) In "Rape of the Lock," the word "lock" refers to a piece/lock of ____.
13.) "Essay on Man" says that, rather than trying to figure out God completely, "the proper study of _______ is man."
14.) In trying to clarify the role of man in the universe, is Pope more involved with optimism or pessimism?
15.) What is the rhyme scheme of the first six lines of a poem written in heroic couplets?
17.) How many *syllables* are in lines written in heroic couplets?
19.) The "thesis" idea of "Essay on Man" is to
"Laugh where we must, be candid where we can;
But ______ the ways of God to man."
22.) Lines written in groups of twos are called _______.

Jonathan Swift

Jonathan Swift Puzzle Clues

Across

1.) The name that "A Modest Proposal" uses for Catholics.

4.) If a Struldbrugg happens to [wed] one of his own kind, "the _____ is dissolved . . . as soon as the younger of the two comes to be fourscore."

5.) On his sea voyages, Gulliver is often employed as the ship's _______.

6.) He goes to Laputa, Balnibarbi, Lagado, and Brobdingnag.

11.) According to "A Modest Proposal," one inhabitant of ______ (name the country) says "that a young healthy childe well nursed is at a year old a most delicious, nourishing, and wholesome food, whether stewed, roasted, baked, or boiled."

13.) The capital city on the lower island of Balnibarbi which contains the academy of projectors.

14.) In the poem "A Description of a City Shower," are the examples lofty or commonplace?

17.) According to the glossary of "Poetic Forms and Literary Terminology" in the back of our textbook, the term _____ "designates literary forms which diminish or derogate a subject by making it ridiculous and by evoking toward it amusement, or scorn, or indignation." Another handbook says that it is "a work or manner that blends a censorious attitude with humor and wit for improving human institutions or humanity. [Practitioners of this approach] attempt through laughter not so much to tear down as to inspire a remodeling. If attackers simply abuse, they are writing invective."

18.) Swift wrote both prose and poetry—true or false?

19.) In "A Description of a Morning Shower," the word "welkin" is an archaic word meaning ____.

21.) Swift characters who don't die but nevertheless get forgetful, opinionated, and peevish, thereby lampooning the idea that living forever is great.

25.) The Houyhnhnms are rational ________ (200 years before the *Planet of the Apes* movies came out).

26.) In his *Drapier Letters* and in other ways, Swift championed the cause of what country subset of Great Britain?

28.) In "A Tale of a Tub," what was the name of the brother who wanted to reduce his outfit to the basics by tearing off all the decorative braid (symbolizing the Calvinists)?

29.) A word etymologically derived from the word "mask," this term signifies a character or identity (such as Gulliver or the speaker/projector of "A Modest Proposal") that an author can hide behind and speak through.

Down

1.) The brother in "A Tale of a Tub" who wants to add more meretricious braid to his outfit, thereby symbolizing Roman Catholicism. (Think of the name of the first pope.)

2.) Swift left the Whig political party to join this other party and write political pamphlets for them, yet this party ultimately fell when Queen Anne died in 1714.

3.) The pejorative name given to men who dream up optimistic but poor ideas, such as the speaker in "A Modest Proposal" or the members of the Academy in Lagado.

7.) The name of the bestial beings driven by base impulses in book IV of *Gulliver's Travels.*

8.) Based on the shapes of the food that they serve, this is the branch of mathematics that the people of Laputa seem to admire most.

9.) These Laputan servants have the job of reminding bright but unrealistic thinkers that it is their turn to listen or to speak.

10.) Because of his low, curmudgeonly estimation of people in *Gulliver's Travels* and elsewhere, Swift was sometimes called a ______, "a hater of humanity."

12.) In book I of *Gulliver's Travels,* the Big-endians and the Little-endians are arguing about which end of an ____ ought to be cracked open first—a satire on the triviality of some political disagreements.

13.) Gulliver's first name.

15.) The floating island propelled by magnetic lodestone forces.

16.) Along with Pope, Swift was one of the two leading literary leaders in the _____ (first or second?) half of the 18th century.

20.) In "A Description of a City Shower," these are the open ditches along the side of a street.

22.) Swift was born in this city and later in life served as dean of St. Patrick's Cathedral here though that was not as prestigious a church post as he had hoped for.

23.) This is the name for a place of Edenic perfection, which is some ways is what the land of the Houyhnhnms is.

24.) The 18th century was a period that produced a number of great sonnets—true or false?

27.) The speaker of "A Modest Proposal" suggests that the Irish ____ children.

Samuel Johnson

Samuel Johnson Puzzle Clues

Across

1.) The Prince of Abyssinia

3.) Johnson has justifiably been criticized for his momentary wordiness when he defined this word in his *Dictionary* as "anything reticulated or decussated, at equal distances, with interstices between the intersections."

4.) A former student of Johnson who came to London with him in 1737, this young man eventually went on to become one of the 18th century's most famous actors.

5.) 1 Across starts off living in the The _____ Valley; its name is ironic because he's not content here but instead wants to leave and see what the world has to offer.

7.) The large city in which Johnson lived and compiled his dictionary; also the name of an early poem of his which served as a harbinger of his literary ability

10.) The conclusion of "The Vanity of Human Wishes" asks, "Where then shall _____ and fear their objects find?"

12.) The female monarch (died in 1714) to whom Johnson was taken to be touched in order to cure the disease in 2 down.

13.) The ancestral Scottish name of Boswell and his father. (See the book's into to Boswell.)

17.) Boswell wrote a journal about a walking tour that he and Johnson took in these English islands.

18.) Johnson published this great work in 1755 after years of work. (It garnered such fame for him that he was later nicknamed ______ Johnson.)

19.) This early essay periodical (1711-1714) by Addison and Steele became the model for Johnson's work in the same genre, such as 1 down.

24.) Johnson said in his "Preface to Shakespeare" that Shakespeare was a great dramatist because "he holds up to his reader a faithful ____ of manners and of life" since his plays reflect the way people really feel and act.

26.) Rasselas's wise guide.

27.) Sir Joshua ______, one of the greatest artists of the 18th century and a member of The Club to which Johnson belonged.

28.) Johnson wrote a great critical biographical series, "The Lives of the _____," in which he detailed the lives of Pope, Milton, Cowley, and others.

29.) The homeland of James Boswell; when he and Johnson first met, Boswell apologized for being from here because Johnson did not typically like people from here. For this verbal exchange, see *The Life of Johnson* section called "A Memorable Year: Boswell meets Johnson."

DOWN

1.) Johnson's publication of this, along with the *Idler* and some *Adventurer* papers were Johnson's contribution to the periodical essay genre; he wrote 208 of these papers.

2.) As a child, Johnson suffered from this pox-like disease, which caused him to be partially deaf and blind on one side.

3.) Rasselas's sister.

5.) "The Vanity of Human Wishes" ends by saying that we should leave to _____ the choice and the amount of the talents and gifts that we receive.

6.) The details of Johnson's life and conversation are preserved for us because this young friend of his wrote a landmark biography of him, and their friendship became widely known.

8.) The name of Johnson's one play, which was not a success.

9.) Johnson self-abasingly defines this vocational person in his *Dictionary* as "a harmless drudge."

11.) According to his "Preface to Shakespeare," this is how long a piece of writing must survive in order to be judged a success and to have passed the test of time.

12.) According to his *Dictionary* and to his "Preface to Shakespeare" (see the footnote), this is what Johnson called a pun.

13.) According to the beginning of "The Vanity of Human Wishes," many people shun fancied ills and pursue _____ goods.

14.) Lord _____ did not cooperate when Johnson sought a wealthy patron for his *Dictionary* project around 1748, eventually prompting Johnson to write a famous letter scolding this peer and refusing his help when it came too late.

15.) Johnson and Boswell were first introduced in a London bookshop owned by this man, a mutual friend of theirs. The location of his bookshop is included in the London map in the endpapers just inside the back cover of our *Norton Anthology*.

16.) Johnson's hometown in central England.

20.) In order to examine all the evidence about life, "The Vanity of Human Wishes" says that it will survey mankind "from China to _____"—an early example of empiricism.

21.) Johnson's pet name for his wife, Elizabeth.

22.) "The Vanity of Human Wishes" says that some people like a young enthusiast are misguided because "Resistless burns the fever for _____."

23.) In his "Preface to Shakespeare," Johnson acknowledges that Shakespeare does not follow the 3 dramatic _____ of time, place, and action which some playwrights felt compelled to follow.

25.) The *Dictionary* humorously defines this as something that Scotsmen eat but that, in England, is only fed to horses.

English Literature I Authors

English Literature I Authors Puzzle

Across

1.) This Renaissance playwright was a contemporary of Shakespeare and wrote *Tamburlaine* and *Dr. Faustus*. Although he died at age 29 in a barroom brawl, during his lifetime he was even one of several writers suspected of authoring plays that we now attribute to Shakespeare.

3.) Author who first wrote a pastoral work but later moved on to write a long, unfinished poem which combined allegory and courtesy book goals with the epic.

6.) Author of *The Life of Johnson.*

9.) Wrote 37 plays and 154 sonnets and considered the best playwright in English.

12.) Leader of the metaphysical poets, this poet was known as a preacher during his lifetime and only achieved fame as a poet after his death in 1631.

13.)Author of *Canterbury Tales.*

17.) Poet who wrote "Elegy Written in a Country Churchyard."

18.) Poet who told readers that time is limited so they ought to gather rosebuds while they can.

19.) Although he wasn't always accepted by other 18th-century novelists, this man wrote *Moll Flanders* and *Robinson Crusoe*.

21.) Author of the epistolary novel *Humphry Clinker* as well as *Peregrine Pickle.*

23.) Wrote a long poem involving Moloch, Chemos, Mulciber, and Beelzebub.

25.) Wrote the unusual 18th-century novel *Tristram Shandy*, which Johnson skeptically described by saying, "Nothing odd will last long."

26.) 18th-century person and acquaintance of Johnson who wrote "The Deserted Village," "The Vicar of Wakefield," and "The Citizen of the World" papers.

28.) This Catholic poet irritated some mature literary critics by writing "Essay on Criticism" when he was only 21 years old.

29.) Poet laureate who wrote the satiric "MacFlecknoe" and helped set the tone for the upcoming 18th century even though he died in 1700.

Down

2.) His "spooky" novel The Castle of Otranto was one of the first to usher in a taste for feeling and emotions which was the harbinger of the Romantic movement in literature which was coming to end the 18th-century ideas of reason, order, and control.

3.) Co-authored *The Tatler* and *The Spectator* with Addison.

4.) Often called the greatest satirist in English, he wrote *Gulliver's Travels* in 1726.

5.) He was the dominant literary figure in English in the second half of the 18th century and wrote poems, a novella, literary criticism, biographies, a play, and a 1755 dictionary.

6.) He wrote the famous allegorical Christian novel *Pilgrim's Progress.*

7.) His female cousin was Henry VIII's fifth wife and he, dying at age 30, didn't live much longer than she did. But during his short life he helped bring the sonnet form to England and was also the first poet to publish a work in blank verse (though this one was not a sonnet since all sonnets rhyme).

8.) Early novelist who wrote *Joseph Andrews* and *Tom Jones.* His sister Sarah even wrote novel titled *David Simple.*

10.) He wrote over a million words, including a famous diary that described the great London fire of 1666 among other things.

11.)Early British novelist who, although he only turned to writing about age 40, wrote the epistolary novel *Pamela* as well as the less happy *Clarissa.*

14.) He helped co-author *The Spectator* after serving political posts in Ireland for many years. He also worked on *The Spectator* in 1714 with Thomas Tickle and also helped with *The Guardian.*

15.) His book which chronicles religious and political martyrs (from those who died under the Catholic persecution of Bloody Mary Tudor to unlucky political victims like Lady Jane Grey) was one of the best selling books in the 17th century after the Bible.

16.) He wrote a poem using the illustration of a ship tossed at sea in a storm to dramatize how he felt in the painful throes of unrequited love.

18.) He wrote only poems of religious devotions.

20.) Although his name sounds like the author of *Rasselas,* this contemporary of Shakespeare spelled his differently and wrote plays like *Every Man in his Humour.*

22.) English philosopher connected to the 17th and 18th centuries wrote *An Essay Concerning Human Understanding* and was one of the few Englishmen really involved with the Enlightenment movement on the continent.

23.) He wrote "To His Coy Mistress."

24.) This Renaissance poet and contemporary of Spenser and Shakespeare wrote sonnets plus a long, narrative, pastoral poem titled "The Countess of Pembroke's Arcadia," often simply shortened to "Arcadia."

27.) Imprisoned and ultimately beheaded for standing up to Henry VIII, this clergyman poet wrote of an idealistic land in which toilets, children's toys, and criminals' chains are made of gold and jewels to show the people's rational contempt for such things which Europeans value highly.

The Romantics

Romantics Puzzle Clues

Across

1.) "Hail to thee _____ Spirit!
Bird that never wert --
That from Heaven, or near it,
Pourest thy full heart
In profuse strains of unpremeditated art.

2.) The first name of Coleridge's wife.

4.) "That time is past,. . . .
And all its dizzy raptures. Not for this
Faint I, _____ mourn nor murmur; other gifts
Have followed; for such loss, I would believe,
Abundant recompense."

7.) "The Rime of the Ancient Mariner" is structurally a ______ narrative because it consists of a story within a story. *The Canterbury Tales* and *The Wizard of Oz* are both _____ narratives too.

8.) The Romantics are most often associated with the _____ District in northern England because Wordsworth and Coleridge moved there in 1799 and wrote poetry amongst its hills and lakes.

10.) The type of bird that the Ancient Mariner shot with his crossbow.

12.) Prior to 1798, Wordsworth lived in this small village in southwest England, and Coleridge lived only a few miles away in Nether Stowey. Dorothy Wordsworth wrote a journal named after this town.

15.) In the last stanza of "Ode on a Grecian Urn," Keats refers to the urn using this adjective meaning Greek.

17.) The name of the large river that runs through London. Wordsworth composed a sonnet upon Westminster Bridge, which spans this famous river, whose name originated from an old Celtic word meaning "ring of bright water."

19.) How many children out of wedlock did Wordsworth have when he spent time in France in the early 1790's?

21.) "A thing of _____ is a joy forever."

24.) The name of the famous prison in Paris that fell when the commoners took over during the French Revolution. This day is still celebrated as a holiday.

29.) A _____ lives in a monastery.

30.) What country had a revolution in 1789 that interested Wordsworth and Coleridge because of its attention to the common man?

32.) Once the Ancient Mariner blessed the sea snakes "unawares," he was immediately able to _____ once again.

33.) An epic poem typically begins in medias ______.

34.) Of the five major Romantic poets, only one of them lived past 1845. Who was it?

35.) PreRaphaelite painters in the 19th century painted beautiful women who often held what flower in their hands to symbolize Christianity?

37.) Give the initials of the author who wrote Biographia Literaria in 1817.

38.) Coleridge and Southey had an idea about moving to Pennsylvania and starting an ideal community which Coleridge called a ________ on the banks of the Susquehanna River.

42.) Shelley's "Adonais" takes the form of a ______ elegy.

44.) Wordsworth said in the great Preface to *Lyrical Ballads* that good poetry was "_____ recollected in tranquility."

45.) Christabel was a _____, which is a creature that is half woman and half serpent.

46.) "My heart aches, and a drowsy _____ pains
My sense, as though of hemlock I had drunk,
Or emptied some dull opiate to the drains
One minute past, and Lethe-wards had sunk."

48.) There were no major Romantic poets who were women. But give the initials of the female Victorian poet who wrote "Sonnets from the Portuguese" later in the century and did achieve considerable fame during her lifetime.

49.) One of two men in the Bible who didn't die but were instead just transported to heaven without dying.

Down

1.) Coleridge's "Rime of the Ancient Mariner" is written in a 4 lines per stanza format called ________ stanza.

2.) The first name of Coleridge's wife.

3.) "Ode to the West Wind" is written in a stanzaic form called terza _____, which was first brought to England in 1557 by Wyatt and Surrey.

4.) In part II of "Rime of the Ancient Mariner," the ship has rounded Cape Horn and is now sailing what compass direction?

5.) In a poem called "The _____ Turned," Wordsworth speaker tells a friend to put down his books and come outside because he can learn more from nature than he can from books.

6.) The young female author of *Frankenstein* had a famous poet brother with the initials ______.

9.) Although his religion was certain heterodox and unusual, Coleridge still sounds pretty conventional in line 470 of "The "Ancient Mariner" when he asks supplicatingly,

"We drifted o'er the harbor bar,
And I with sobs did pray—
O let me be awake, my _____
Or let me sleep always."

11.) A lamia is half woman and half _____.

13.) In the conversation poem "Frost at Midnight," Coleridge's speaker mentions a fragile film called a "stranger" which flutters as he stares at the _____.

14.) "Ozymandias" is another name for the man who was once ruler of what north African country?

16.) Keats lost his mother and his younger brother Tom to the disease _____, which eventually killed him as well at age 25.

18.) Unrhymed iambic pentameter is also called _____ verse.

20.) Did Lord Byron live longer than Coleridge?

22.) What is the rhyme scheme for the first 3 lines of an English sonnet?

23.) According to Wordsworth, which phase of life is more attentive and capable of absorbing life---youth or old age?

25.) Despite mental illness, Dorothy Wordsworth lived until 1855, a year made famous in America as the year that Whitman published ______ *of Grass* and Melville published *Moby Dick.*

26.) What was the name of the sacred river in Xanadu where Kubla Khan built his stately pleasure dome?

27.) This man was a good friend of Coleridge whose wife was the sister of Coleridge's wife.

28.) A foot of poetry with 2 syllables, and the first one is unaccented while the second one is accented.

29.) "In nature and the language of the sense
The anchor of my purest thoughts, the nurse,
The guide, the guardian of my heart, and soul
Of all my ______ being."

31.) One of the first rumblings of the Romantic movement in the late 18th century was the popularity of ______ novels such as Walpole's *The Castle of Otranto* and Matthew Lewis's *The Monk*, which put a new emphasis on feelings and emotions such as horror.

34.) Wordworth watershed poem "Tintern Abbey" was written in the valley of the ____ River, which Wordsworth had visited 5 years earlier.

36.) "I ______ lonely as a cloud
That floats on high o'er vales and hills,
When all at once I saw a crowd,
A host, of golden daffodils."

37.) A term for anything that is huge, intimidating, awe-inspiring, lofty, and dangerous. Wordsworth liked the Alps or storms for this reason.

39.) A figure of speech in which a number of poetic lines or a number of sentences all start with the same word or words.

40.) Keats was famous for writing this type of poem which exalts an object, a season, or a mood. This sub-genre of poetry comes in Pindaric and Horatian types.

41.) "Lines Written in Early Spring" concludes with the quatrain

" If this belief from heaven be sent,
If such be Nature's holy plan,
Have I not reason to lament
What _____ has made of _____?"

42.) Wordsworth's poem "Elegaic Stanzas" came from an idea he got from a picture of ______ Castle painted by Sir George Beaumont.

43.) Although she does not share the poetic agenda common to the Romantic poets, novelist _____ Austen, who wrote Emma and Pride and Prejudice nevertheless lived in the same chronological period that Wordsworth, Coleridge, and the others did.

46.) The 3 initials (including her maiden name) of the young woman who wrote *Frankenstein.* She had a well-known mother and a well-known husband.

Wordsworth

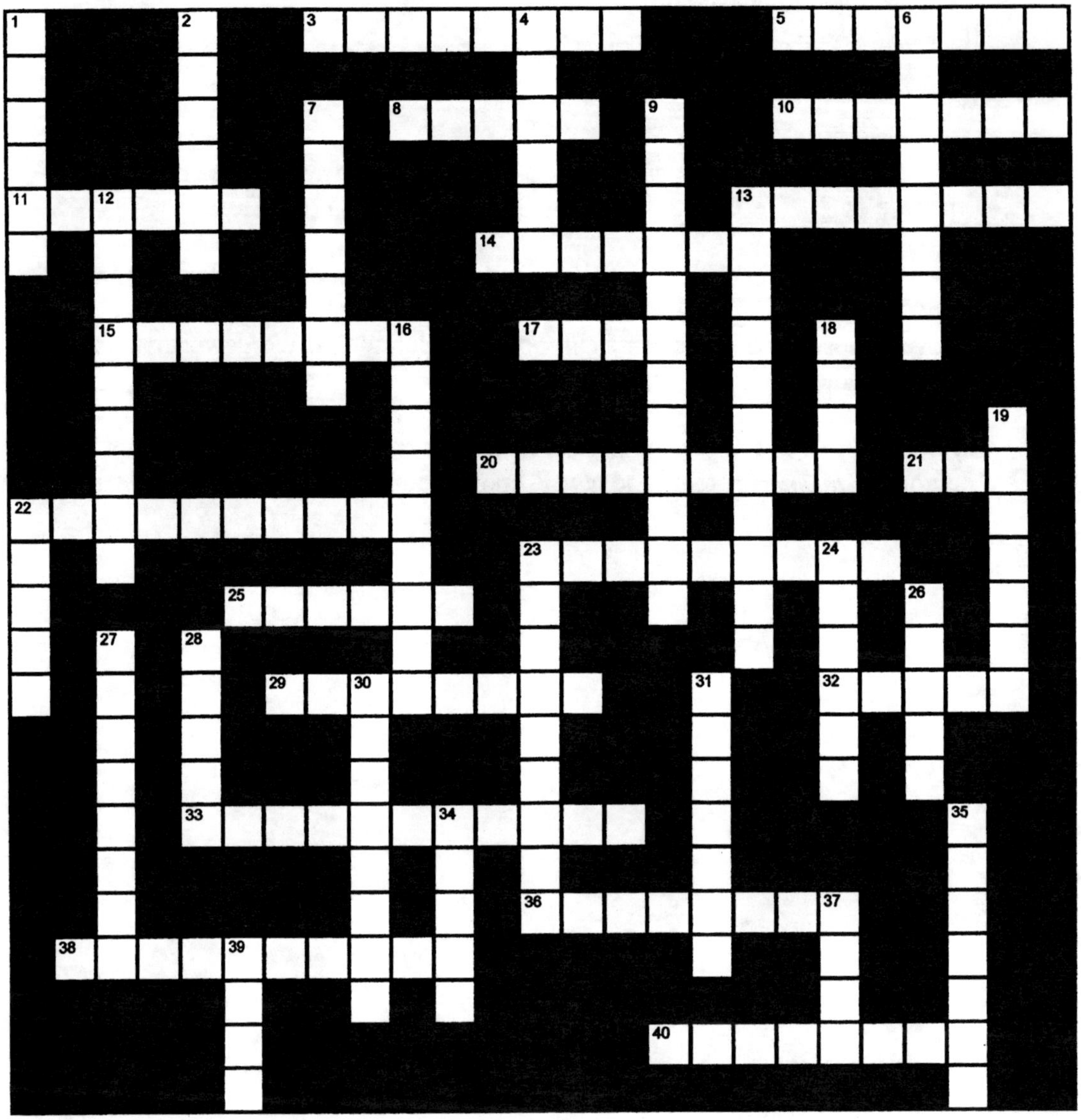

William Wordsworth Puzzle Clues

Across

3.) In 1843 Wordsworth was appointed poet _____ of England, the highest official governmental position for a poet although some people, like Browning, viewed this as the young rebel poet Wordsworth selling out to the status quo establishment.

5.) Compared to 18th-century poets, which was Wordsworth most interested in: discipline, order, passion, or social conformity?

8.) In book 4, lines 298-362 of *The Prelude*, Wordsworth experiences an epiphany on the way home from a _____, and it becomes clear to him that poetry will be his life's work rather than the superficiality of social events.

10.) A word (which he may has obtained from Edmund Burke or Longinus) meaning "characterized by nobility, grandeur, majesty; inspiring awe and supremely impressive and moving"; but also large with a touch of danger and mystery. He found the towering Alps mixed with plunging crevices to be this, which is why he toured the Alps and why they affected the feelings that he considered connected to great poetry.

11.) Was Wordsworth (WW) more interested in writing poetry about people who were common or people who were famous?

13.) How old was WW when his parents died?

14.) In "Elegiac Stanzas" WW suggests that he has lost some of his poetic vigor and inspiration as he says,

> So once it would have been—'tis so no more;
> I have submitted to a new control:
> A power is gone, which nothing can restore;
> A deep distress hath humanized my Soul."

In saying this, he is alluding to the drowning death of his ______.

15.) Early in the 1790's, WW and Coleridge lived near each other in southwest England; Coleridge lived in Nether Stowey, and WW lived about 3 or 4 miles away in _____, from which Dorothy wrote some journals that still exist.

17.) "Resolution and Independence" ends with the speaker saying, "I'll think of the Leech Gatherer on the lonely ____." This word is a vast, desolate, flat type of land with grasses and heath, which also functions as a prominent setting in Act III of Shakespeare's *King Lear* and in the Sherlock Holmes story, "The Hound of the Baskervilles."

20.) "I wandered lonely as a cloud,
That floats on high o'er vales and hills,
When all at once I saw a crowd,
A host, of golden ______
Beside the lake, beneath the trees."

21.) The great watershed poem "Tintern Abbey" was composed (as its subtitled tell us) near the banks of this English river, which WW was revisiting after an absence of 5 years.

22.) Of the 5 major Romantic poets (Wordsworth, Coleridge, Byron, Shelley, and Keats), which one lived the longest?

23.) Of the 5 major Romantic poets, which one lived second longest?

25.) As opposed to the philosophy of the Romantic poets, did the 18th-century poets prefer to live amongst the hillsides or in cities?

29.) WW and Coleridge were quite interested in the French Revolution, which began with the storming of the ______, the terrible prison in Paris.

32.) The word "sylvan" refers to the ______.

33.) WW felt that poetry was "the _____ overflow of powerful feelings."

36.) While in France in 1792, Wordsworth fell in love with Annette Vallon, and the 2 of them had a _____ named Caroline.

38.) In "Tintern Abbey" WW admits that he has lost some of the freshness and spontaneity of youth, but this has been offset by other abilities:

"Not for this
Faint I, nor mourn nor murmur; other gifts
Have followed; for such loss, I would believe,
Abundant __________. "

39.) How many lines are there in each of WW's sonnets, "Surprised by Joy" and "The World is Too Much with Us"?

Down

1.) What country did Wordsworth go to in 1791 and 1792 because he admired the groundswell of political action by the common people?

2.) While in the country in 1 down, he fell in love with a woman named Annette ____.

4.) "Therefore am I still
A lover of the meadows and the woods,
And mountains; and of all that we behold . . .
In nature and the language of the sense
The _____ of my purest thoughts, the nurse,
The guide, the guardian of my heart, and soul
Of all my moral being."

6.) Did Wordsworth and Coleridge prefer solitude or the company of lots of people?

7.) This poem, originally begun in two parts in 1799, is WW's longest poem and was published upon his death in 1850. The Norton Anthology calls it "a personal history which turns on a mental crisis and recovery [much like] a spiritual autobiography of crisis."

9.) WW's close friend Coleridge lived for several years in this town in southwest England before the 2 men moved north to the Lake District.

12.) The site of Tintern Abbey was itself the ruins of a Catholic _______ , which King Henry VIII had torn down when England broke away from Roman Catholicism in 1539.

13.) According to the important Preface to *Lyrical Ballads*, WW felt that good poetry was "emotion recollected in _______."

16.) The Lake District, which became the central location for the early Romantic poets, is located in which area of England—southwest, east, west, northwest?

18.) WW traveled to visit these soaring mountains in Germany.

19.) The solitary old man in "Resolution and Independence" makes a modest living by gathering _____.

22.) A famous line from Browning says that "God's in his heaven, all's right with the _____."

23.) "Tintern Abbey " says that this period of life is the most fresh, keen, and receptive though animal-like in not being able to understand or interpret every thought.

24.) The poem in 7 down is subtitled "The ____ of a Poet's Mind" because it narrates and chronicles Wordsworth's increasing interest in poetry.

26.) In "The Tables Turned" WW tells a person to leave his _____ alone to keep from becoming doubled over by the arduousness of studying; besides

"_____! 'tis a dull and endless strife:
Come, hear the woodland linnet,
How sweet his music! On my life,
There's more of wisdom in it."

27.) When in the Lake District, Wordsworth live in the town of _______ in a small house that we now call Dove Cottage, and his sister kept journals of their lives there.

28.) Wordsworth created the term "_____ of time" to designate those moments in life in which we see truth or the meaning of the past most clearly.

30.) In "The Solitary Reaper," the speaker can't understand the words the words that the solitary young woman is singing because she is singing in Erse, which is a version of Gaelic spoken in the country of _______.

31.) The name of Wordsworth's sister, who was very close to him and often lived with him until he got married.

34.) "I heard a thousand blended ______
While in a grove I sate reclined,
In that sweet mood when pleasant thoughts
Bring sad thoughts to the mind."

35.) WW said in the Preface to *Lyrical Ballads* that great poetry was "the spontaneous overflow of powerful _____."

37.) In the poem "Simon Lee" the speaker stops to help an old man who isn't able to cut the _____ of a tree.

39.) The first name of Wordsworth's wife.

The Victorians

Victorians Puzzle Clues

Across

1.) *Wuthering Heights* was written by ______ Bronte in 1848; "Wuthering" is a Yorkshire regionalism for "weathering," meaning "stormy."

3.) *Middlemarch* is a novel written by George _______, the penname for Mary Ann Evans.

5.) William ______ , a man of varied talents, operated the Kelmscott Press, designed fabric with oak leaf patterns, furniture, and played a role in the arts and crafts movement.

6.) *Tess of the D'Urbervilles* was written by Thomas _____, who also wrote poems such as "The Darkling Thrush."

8.) When Robert Browning and Elizabeth Barret eloped, they fled to Italy and settled in the city of _______ in a house called Casa Guidi.

11.) John Henry Cardinal Newman wrote a work called *Apologia Pro ____ Sua,* which translates as *A Defense of My [Religious] Life.*

13.) As opposed to the novelists Thackeray and Dickens, men such as Tennyson, Browning, Wordsworth, and Arnold all were _____.

14.) Queen Victoria married Prince _______, who organized the Royal "Crystal Palace" Exhibition of 1851 but lived decades shorter than his wife.

16.) The women in PreRaphaelite paintings are often holding what flower that has a religious significance?

17.) At the beginning of Dickens's novel *Hard Times*, Mr. Gradgrind insists, "Now, what I want is, _____. Teach these boys and girls nothing but _____. ____ alone are wanted in life. Plant nothing else, and root out everything else . . . . Nothing else will ever be of any service to them."

20.) In Browning's "The Bishop Orders his Tomb in St. Praxed's Church," the bishop wants his tomb adorned by a lump of this blue precious stone, _____ lazuli.

21.) A foot of poetry which has two syllables with the first one accented and the second unaccented is called a ______ foot.

22.) The last name of the young college friend whose death caused Tennyson to write the elegy "In Memoriam, A.H.H."

23.) The Victorian poet most famous for his dramatic monologues.

25.) "Half a ____, Half a ______
Half a ______ onward,
Into the valley death
Rode the six hundred."

27.) This minor novelist wrote a successful supernatural story, *The Woman in White,* and one of the first detective novels, *The Moonstone.*

29.) Matthew Arnold wrote the poem "The _______ Life" about deep, warm feelings that live within us but ones that we only tap into with the person that we love the most.

30.) The nickname for Robert Weideman Browning, who was the son of Robert and Elizabeth Barrett Browning.

32.) Near the bottom of his staircase, Browning's homicidal Duke has a statue of this Roman god taming a seahorse.

33.) Browning wrote a poem about a horseback rider galloping headlong in an effort to carry a message of peace from the city of Ghent to _______.

34.) The longest river in the world.

35.) In "The Second Coming," Yeats writes of a monster slowly lumbering toward the Hebrew city of ________, ushering in 2,000 years of terrible conditions.

38.) In "Prospice," Browning speaks of his own death and says that he will fight one fight more and that it will be his last and his _____.

39.) Eliot's novel *Silas Marner* is about a man who is a ______ by trade, as the subtitle says.

40.) Anthony Trollope is sometimes called the first English writer of serial fiction because he wrote 6 novels, such as *The Warden* and *Barchester Towers*, about people living in the rural county of ________.

Down

2.) What is the name of the turn of the century Irish poet who wrote "The Swans of Coole" and "Sailing to Byzantium," plus other poems involving Irish legends?

4.) The Tractarian Movement involving John Henry Newman and others centered around what major English university?

5.) One of Browning's most successful poetry collections came out in 1855 and was titled _______ *and Women.*

7.) Any poem in the same subgenre as Arnold's "Adonais" is written about a friend who is ______.

9.) "Adonais," like Milton's "Lycidas," belongs to a subgenre called a pastoral _____.

10.) In 1859, Charles Darwin produced *On The Origin of the Species*, a book about ________, a book which sold out its first edition the same day it went on sale.

11.) Browning's bishop who is ordering his tomb wants an epitaph in good Latin written by the Roman poet _______, also known as Cicero.

15.) In a poem bearing her name, Porphyria is quietly killed by her lover using her own _____ to strangle her.

17.) Browning became an atheist in his early adult years after reading Shelley, but later he regained his ______ in a Christian God and wrote a number of poems about clergymen, some good and some bad.

18.) What are the initials of the woman who wrote "Goblin Market" and whose poet/painter brother wrote "The Blessed Damozel" and helped form the PreRaphaelite Brotherhood of painters?

19.) What was the name of the artistic group of Cambridge undergraduates who befriended and influenced Tennyson so much?

20.) The Lady of Shalott "left the web, she left the _____
She made three paces round the room,
She saw the water lily bloom."

23.) In Tennyson's "The Passing of Arthur," to which young knight did Arthur entrust the job of returning the sword Excalibur to the Lady of the Lake?

24.) The last poem in any collection of Browning's poems is the "Epilogue to Asolando," a poem in which Browning speaks of his death and of courage. According to the poem, at what time of day "in the bustle of man's work-time" should we "Greet the unseen with a cheer"?

26.) The man who is charge of a monastery is called an _____.

27.) The initials of the man who wrote *Bleak House*.

28.) In 1850, Tennyson was appointed poet _____ of England, replacing Wordsworth, who had just died.

31.) The pseudonym of Mary Ann ____ was George Eliot.

34.) Did Elizabeth Barrett Browning outlive her husband?

36.) The initials of the English writer who translated the Persian poetic work, *The Rubaiyat of Omar Khayyam*. He later insulted Mrs. Browning after she died, and Browning wrote a poem which was a stinging response.

37.) What is the rhyme scheme of the first three lines of an English sonnet?

38.) Robert Browning was ____ in 1812 one year after Thackeray and three years after Tennyson.

40.) What were the first two initials of the turn-of-the-century English novelist who wrote *A Passage to India* and *Howard's End*?

Alfred Lord Tennyson

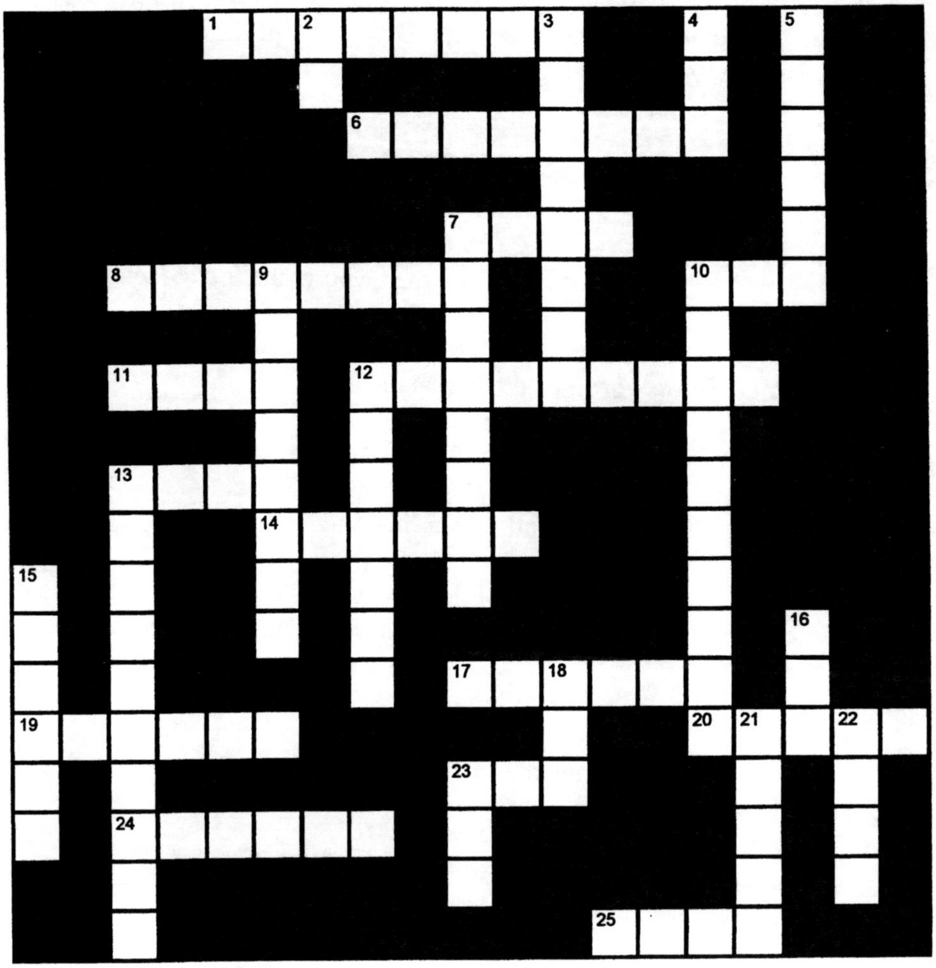

Tennyson Puzzle Clues

Across

1.) Who is the knight whom the Lady of Shalott sees riding past her tower window?

6.) What was the name of the group of Cambridge undergraduates interested in the arts who influenced Tennyson and encouraged him to continue to write verse?

7.) "She left the web, she left the _____,
She made three paces round the room,
She saw the water lily bloom,
She saw the helmet and the plume."

8.) She was queen of England for an amazing 64 years from 1837 to 1901.

10.) The last Tennyson poem in any collection of his poems (including our textbook) should be "Crossing the ____," which refers dying.

11.) "In Memoriam" mentions one reason why Tennyson found solace in writing the long poem about the death of his close friend:

"But, for the unquiet heart and brain,
A use in measured language lies;
The sad mechanic exercise,
Like dull narcotics, numbing _____."

12.) Time in "In Memoriam" is subdivided by having three ______seasons break up the poem into sections which show the poet gradually healing from the loss of his friend.

13.) "Ulysses" and "The Lotus-Eaters" are both poems that refer to Homer's Odyssey and to Ulysses's long trip home after fighting in the 10-year war at _____.

14.) When The Lady of the Lake gave Arthur the sword Excalibur, her arm was clothed in glittering _______ cloth.

17.) "______ of the King" is a sequence of Arthurian poems that Tennyson wrote.

19.) "Mariana" is a poem derived from the line "Mariana in the Moated ______," which comes from Shakespeare's play *Measure for Measure*.

20.) A type of poem written to commemorate a dead person. It can sometimes be pastoral, but "In Memoriam" is not completely pastoral.

23.) Ring out the old, ring in the _____,
Ring happy bells, across the snow:
The year is going, let him go;
Ring out the false, ring in the new.

24.) The name of the Tennyson's close Cambridge friend whose death at 22 had a big effect on Tennyson and was the subject which caused Tennyson to write "In Memoriam" over a period of years.

25.) "She only said, 'My life is dreary,
He cometh not," she said;
She said, "I am aweary, aweary,
I would that I were _____."

Down

2.) Did Tennyson live to reach his 90th birthday?

3.) Name the Tennyson poem that spoofs the idea that eternal life is great by depicting a person who is very old and hates it.

4.) Were Tennyson, Robert Browning, and Matthew Arnold the 3 major poets of the English Victorian era?

5.) The Lady of Shalott's initial problem is a curse prevents her from experiencing life directly, so she can only view life indirectly by looking at it through a _____, which cracks from side to side when she finally decides this isn't much of a life.

7.) In 1850 Tennyson succeeded Wordsworth as official poet ______ of England when Wordsworth died.

9.) Of Tennyson, Browning, and Arnold, who lived the longest?

10.) What is another name for unrhymed iambic pentameter?

12.) "The Charge of the Light Brigade" was written about an episode in the _____ War, which is the same war in which Florence Nightingale started the study of nursing.

13.) _____ was the son of Ulysses though the two can't have known each other very well because Ulysses had been out of the country for 20 years while his son was growing up.

15.) "Half a ____, half a ____,
Half a _____ onward,
All in the valley of Death
Rode the six hundred."

16.) Tithonus's problem is that he can't _____.

18.) A _____ tree is a tree that is sometimes associated with mourning and graveyards. It appears in the first few pages of "In Memoriam."

21.) One of most famous lines in Tennyson's work is the one that concludes this quatrain:
"I hold it true, whate'er befall;
I feel it, when I sorrow most;
'Tis better to have ______ and lost
Than never to have _____ at all."

22.) "Old age hath yet his honor and his toil
Death closes all; but something ere the end,
Some work of noble note, may yet be done,
Not unbecoming men that strove with _____."

23.) Ulysses laments that
"Matched with an aged wife, I mete and dole
Unequal laws unto a savage race.
That hoard, and sleep, and feed, and know ____ me."

Robert Browning

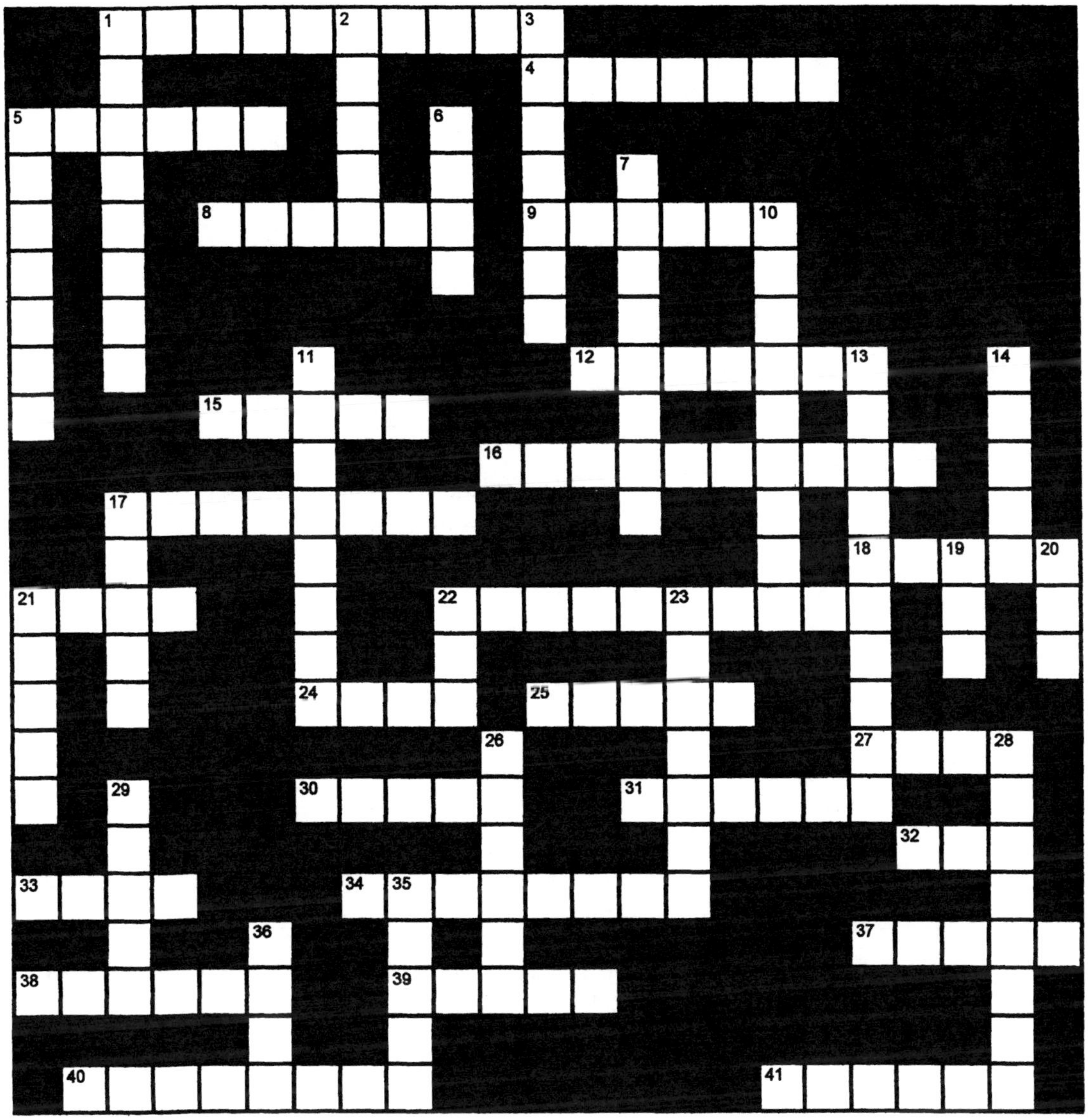

Robert Browning Puzzle Clues

Across

1.) The famous line "How do I love Thee, Let me Count the Ways?" comes from Elizabeth Barrett Browning's collection "Sonnets from the _____."

4.) At the bottom of his staircase, Browning's Duke (in "My Last Duchess") has a statue of ____taming a seahorse, symbolizing the Duke's controlling nature toward one of his creations.

5.) In "The Laboratory," a woman commissions a chemist to prepare some _____ to give to her boyfriend and his other girlfriend on the side.

8.) Fra Lippo Lippi is rightly confident because his patron is the powerful Cosimo of the famous ____ family in early Renaissance Italy.

9.) The bishop who is ordering his tomb has a son named ____ who is holding back, implying suspicion toward his father.

12.) The bishop of St. Praxed's competed with this rival for the love of a woman and still competes with him for a tomb with the nicest position and materials.

15.) A Browning poem in which he says that he will not reveal personal details about his own life, saying that to do so would be like living in a glass edifice that everyone could look in on.

16.) The name of the only Romantic poet who lived to 1850 and eventually became poet laureate of England. But by accepting the poet laureateship, this poet, who had originally been seen as a poetic rebel and model, was considered by Browning to have sold out to the establishment, thereby precipitating the criticizing poem "The Lost Leader," which likens him to Judas Iscariot.

17.) "--Good saints, how if feared
To set my foot upon a dead man's cheek . . .
--It may have been a _______ I speared,
But, ugh! It sounded like a baby's shriek.

18.) An artist says, "A man's reach should exceed his ____ else what's a heaven for?"

21.) A soliloquy poem begins and ends with this guttural growl (skip the dashes).

22.) "She had
A heart—how shall I say?—too soon made glad,
Too easily impressed; she liked whate'er

She looked on, and her looks went _____."

24.) What is the rhyme scheme for the first 4 lines of an English sonnet?

25.) The monk speaker in "Soliloquy of the Spanish Cloister" is going to stick in a page from a dirty French novel when his enemy gathers greengages. A greengage is a type of _____.

27.) One of the Romantic poets that Browning admired most was Percy Shelley, whom Browning called "sun-treader" and who wrote a famous poem "To a Sky- _____."

30.) In 1855, Browning published a collection of poems titled *Men and* _____, which the Norton Anthology once called "his most memorable volume of poems."

31.) "Childe ____ to the Dark Tower Came"

32.) We see into a bishop's religious impurity when he juxtaposes pagan tomb images with Christian ones, such as "one ____/Ready to twitch the Nymph's last garment off,/And Moses with the tables ."

33.) The Norton text says that "his greatest single work, *The Ring and the* _____, a poem on the vast scale of a long novel, was published in 1868 and was based on an actual murder trial concerning suspected marital infidelity.

34.) "G-r-r---there go, my heart's abhorrence!
Water your damned flowerpots, do!
If hate killed men, Brother _____,
God's blood, would not mine kill you!"

37.) The dying bishop of St. Praxed's wants a nice epitaph written by this Roman writer, also named Cicero.

38.) One of Browning's key sources about the lives of actual Renaissance painters like Fra Lippo Lippi and Andrea del Sarto was a book titled *Lives of the Painters* by George _____.

39.) After Robert Browning fell in love with Elizabeth Barrett, who was more famous than he was at that time, they eventually eloped to the country of _____ in order to escape her controlling father and to improve her health in the dry climate there.

40.) When living in the country mentioned in 39 across, they lived in a house called Casa Guidi in the city of _______.

41.) This university in Texas is home to the Armstrong-______ Library, which contains one of the best collections of Browning books and artifacts in the world: letters, furniture from their home, paintings by their son, a lock of her hair, and stained glass windows commemorating his poems.

Down

1.) Andrea del Sarto, Fra Lippo Lippi, the fictitious Fra Pandolf, and even the Brownings' son Pen all shared the vocation of being _______.

2.) Casa _____ was the name of the Brownings' home in Florence.

3.) The country that is home to Bath, York, and Canterbury.

5.)The name of the imaginary artist who painted the picture of "My Last Duchess"

6.) Her lover killed Porphyria by strangling her with her _____.

7.) In any collection of Browning's poetry (as in our textbook too), the last poem of his is always the "Epilogue to _____," published on the day he died.

10.) Browning is widely known for using and popularizing the genre of the dramatic ______, which includes "My Last Duchess," "Andrea del Sarto," and Fra Lippo Lippi."

11.) In the poem bearing his name, Andrea del Sarto has been arguing with his wife, _____, who is apparently listening to hear the whistle signal from her extramarital lover, whom they euphemistically call her cousin.

13.) The surname of the person whom Browning angrily addresses this way after the other writer made insulting comments about Mrs. Browning:

"Kicking you seems the common lot of curs—
While more appropriate greeting lends you grace,
Surely to spit there glorifies your face—
Spitting from lips once sanctified by hers."

14.) _____ lazuli, a valuable blue stone mentioned in "The Bishop Orders his Tomb."

17.) A famous line from "Pippa Passes" says, "God's in his heaven, all's right with the _____."

19.) Onc rhythmic Browning poem narrates the story of a horse rider's headlong journey to take some good news of peace from the towns of Ghent to ____.

20.) Robert and Elizabeth had a son named Robert Weideman Browning, but they just called him by the nickname of _____.

21.) After telling the story of his life and discussing the hardships of a monk, Lippo Lippi says

Well, well, there's my life, in short,
And so the thing has gone on ever since.
"I'm grown a man no doubt, I've broken bounds:
You should not take a fellow eight years old
And make him swear to never kiss the _____."

22.) The 3 initials of Browning's poetic wife.

23.) The Brownings were married at Marlybone Church, which is located in north London on ____ Street. A Broadway play and even a movie about their courtship are titled using this same street name, "The Barretts of _____ Street."

26.) ____ del Sarto

28.) The title of the eerie, quest-like poem "Childe Roland to the Dark Tower Came" is derived from Shakespeare's play ________.

29.) "_____, what's to blame? You thing you see a monk!
What, 'tis past midnight, and you go the rounds,
And here you catch me at an alley's end
Where sportive ladies leave their doors ajar?"

35.) At the end of "Porphyria's Lover," Porphyria is no longer ______.

36.) In "My Last Duchess," the Duke speaker is talking to the envoy of a Count as the Duke tries to make arrangements to get a new _____ to replace the one that he's done away with.

Charles Dickens

Dickens Puzzle Clues

Across

1.) A young law clerk in *Bleak House* who likes Esther Summerson but later tried to wiggle out of his offer of marriage after illness ruins her beauty; he also feels a sense of power when he learns Mrs. Dedlock's secret

4.) 7 down in *Bleak House* keeps a collection of pet _____ named after the traits a of Chancery: Despair, Hope, Joy, Youth, , Want, Ruin, and Madness.

6.) Oliver Twist lands in a workhouse because he is an _______.

7.) *A Tale of Two Cities* takes place both in England and in _____.

8.) The surname of the woman who knows the secret to Esther Summerson's identify, unbeknownst to her elderly patrician husband.

10.) A law clerk in *Bleak House* whose name sounds more like a large container for mop water.

12.) Aunt Betsey's pet name for Copperfield.

14.) The early nanny in *David Copperfield* who gives David refuge and ends up living in a house shaped like a boat.

15.) The cruel schoolmaster in *Nicholas Nickleby* who rules over Dotheboy's Hall.

18.) The irresponsible gentleman in *David Copperfield* who lives off the generosity of others and says that they are beholden to him for using his childlikeness to bring out their altruism.

20.) *A Tale of Two Cities* often focuses on Paris soon after the ______ there in 1789 when the commoners struck back against the haughty aristocrats.

21.) The poor, pitiful boy who sweeps the street corners in *Bleak House* but whose testimony carries no credibility because he's considered nobody.

22.) *The* ______ *Papers* was one of Dickens's first novels about a comical group of club members who travel episodically from place to place.

23.) The lead character in *Bleak House* who suffers from a terrible illness for awhile but is one of the narrators in the novel's dual narrative.

24.) The young heroine in *The Old Curiosity Shop* who was so popular with 19th-century readers.

25.) To construct a metaphor for the murkiness of the English chancery court system, the second paragraph of *Bleak House* uses this weather feature, saying that it is everywhere—on the river and marshes, in eyes and throats, on the cabooses and small boats

27.) Mr. Jarndyce tells Esther that he retreats to The _______ when he gets in a bad mood, i.e., when the wind is blowing in the wrong direction..

28.) The name of the debtor's prison which provides part of the setting in *Little Dorrit.*

29.) Bill Sykes ultimately kills this young barmaid in *Oliver Twist*, which is the beginning of the end for him as everyone starts to hunt him down for his brutal crime.

Down

2.) Fagin teaches his motley collection of boys in *Oliver Twist* to make a larcenous living on the streets by becoming _______.

3.) Ada Clare and Richard Carstone are known as the ______ in Jarndyce.

4.) In addition to 22 across, one of Dickens' earlier writing collections was *Sketches by* _____.

5.) The name of the horrible boys school in rural Yorkshire where Nicholas Nickleby becomes a young schoolteacher.

7.) The frail, mad, elderly woman in *Bleak House* who owns the animals of 4 across and attends court each day to wait for its never-ending verdict.

9.) *Bleak House* is a dark satire of the circumlocution and lethargy of the English _____ system.

11.) The belittled clerk employee of Ebenezer Scrooge in *A Christmas Carol* who is friends with Tiny Tim.

13.) Because its young protagonist has no father figure, works in a factory, and later becomes a writer, David _____ is perhaps Dickens's most autobiographical novel.

15.) The student friend of Copperfield who refers to him as Daisy but later reveals a more negative side.

16.) The manipulative lawyer in *Bleak House* who discovers the secret of Esther's parentage and tries to use it for blackmail.

17.) Many of Dickens's grimy urban novels are set in this large capital city.

19.) Acquaintance of Fagin who is the most brutal and dangerous character in *Oliver Twist.*

21.) The last name of the major lawsuit in *Bleak House*.

25.) The colorful leader of the thieving boys who harbors Oliver Twist for a time but does him no good later on.

26.) Uriah _____ is the slimy, obsequious youthful villain in *David Copperfield.*

The Nineteenth-Century Novel

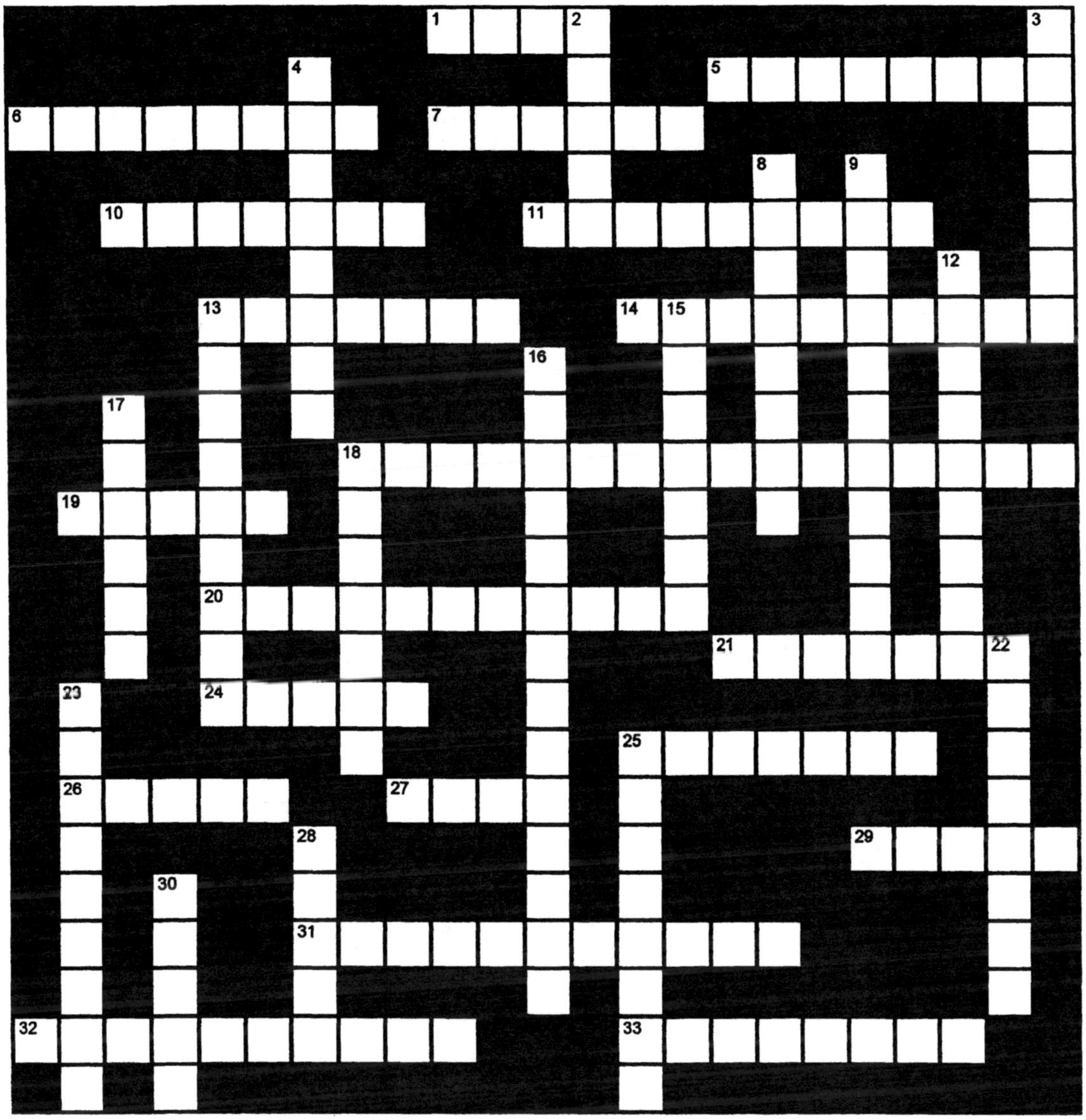

19th-Century Novel Puzzle Clues

Across

1.) Little _____, the heroine of Dickens's *Little Curiosity Shop.* When Dickens wanted to kill off the popular little girl, the public outcry against it was so great that it caused him to let her live.

5.) *Nicholas* _____ is Dickens's novel about young man Nicholas, his sister Kate, and their elderly mother who have to make their way in the world when their father an husband dies and leaves them penniless. The meet colorful characters along the way like good Smike and Newman Noggs as well as their cold Uncle Ralph.

6.) Miss _____ is the old spinster in Great Expectations who manipulates Pip and others out of bitterness for being abandoned at the marriage altar years earlier.

7.) The _____ is the first novel in Anthony Trollope's six-novel series, The Barsetshire Novels, which The Oxford University Press called the first serial fiction in English literature.

10.) Mr. ______ is the highly ethical hero of 7 across who works as a music director and overseer of a group of elderly men living on the will of a philanthropist.

11.) A shorter novel in which Dickens uses Mr. Gradgrind and Mr. M'Choakmchild to satirize the shallowness of utilitarianism.

13.) The foolish vicar and relative in *Pride and Prejudice* who stands to inherit Longbourne when Elizabeth Bennett's father dies.

14.) Possibly Dickens's most powerful novel about Esther Summerson, the wards of Jarndyce, and the maddening chancery court system.

15.) Possibly the most biographical of Dickens's novels, this one involves a boy with a terrible stepfather, a nurturing Aunt Betsey, and an manipulative boy named Uriah Heep.

19.) *Wuthering Heights* was written by _____ Bronte.

20.) The villainous attorney in Bleak House who learns Lady Dedlock's secret.

21.) "Far from the ______ Crowd," Hardy's more optimistic novel about a young female farm owner upon whom more than one man has set his sights.

24.) George Eliot is actually the pseudonym for Mary Ann ______.

25.) The proud, powerful, controlling bishop's wife in Anthony Trollope's *Barchester Towers.*

26.) In Bronte's Jane Eyre, Mr. Rochester keeps an insane woman locked up in the ______.

27.) Hardy's novel of love and murder, ____ *of the D'Urbervilles.*

29.) The 19th-century author of romantic novels such as *The Lady of the Lake* and *Quinten Durward.* Mark Twain didn't like this writer's romanticism, which is why Twain named a sinking riverboat after him in *Huckleberry Finn.*

31.) The Dickens novel which has been remade and retold probably more than any other of his. The plot revolves around an orphan boy whose mother dies giving birth to him, but after many hardships, such as being apprenticed to a mortician, he is eventually found by a kindly gentleman who once knew his father.

32.) Trollope's ______ *Towers*, the second of the novels about the county of Barset and arguably Trollope's best novel.

34.) The scholarly but dry clergyman who marries Dorothea but spends his life futilely writing *The Key to a*ll Mythologies in Eliot's *Middlemarch.*

Down

2.) The foolish Bennett daughter in *Pride and Prejudice* who runs off with Mr. Wickham, causing the family much anguish and, it is feared, possibly ruining the marriage prospects for the other daughter as well.

3.) The idealistic young doctor in *Middlemarch.*

4.) The young, idealistic writer whom no one but Dorothea cares for in *Middlemarch.*

8.) The poverty-stricken but eternally optimistic gentleman in David Copperfield who always retorts that "something will turn up."

9.) The location of an important ball in *Pride and Prejudice*

12.) Surname of Esther in *Bleak House*, who is one of the novel's 2 narrators.

13.) This Bronte sister wrote *Jane Eyre*.

15.) Small town near Darcy's familial estate where Elizabeth's aunt grew up and to which Darcy used to run when a boy.

16.) The last of Thomas Hardy's novels, which received such a negative reaction that he never wrote another novel, turning to poetry instead.

17.) The bumbling parish employee who roughly watched over Oliver Twist and the rest of the workhouse boys.

18.) Enormously popular author of *David Copperfield, Great Expectations, A Tale of Two* Cities who even came to America and wrote *American Notes* that weren't very complimentary of America.

22.) The name of the father-in-law of Septimus Harding and the archbishop in *Barchester Towers* who opposes the Proudies.

23.) Author of *Vanity Fair* and *Barry Lyndon.*

24.) The episodic novel _____ *Papers* was one of Dickens's first, popular, comic novels.

28.) *The Mill on the ______,* a somewhat tragic George Eliot novel.

30.) The wealthy but prejudiced hero of Jane Austen's *Pride and Prejudice* who finally comes to see the error of his ways.

English Literature II Authors

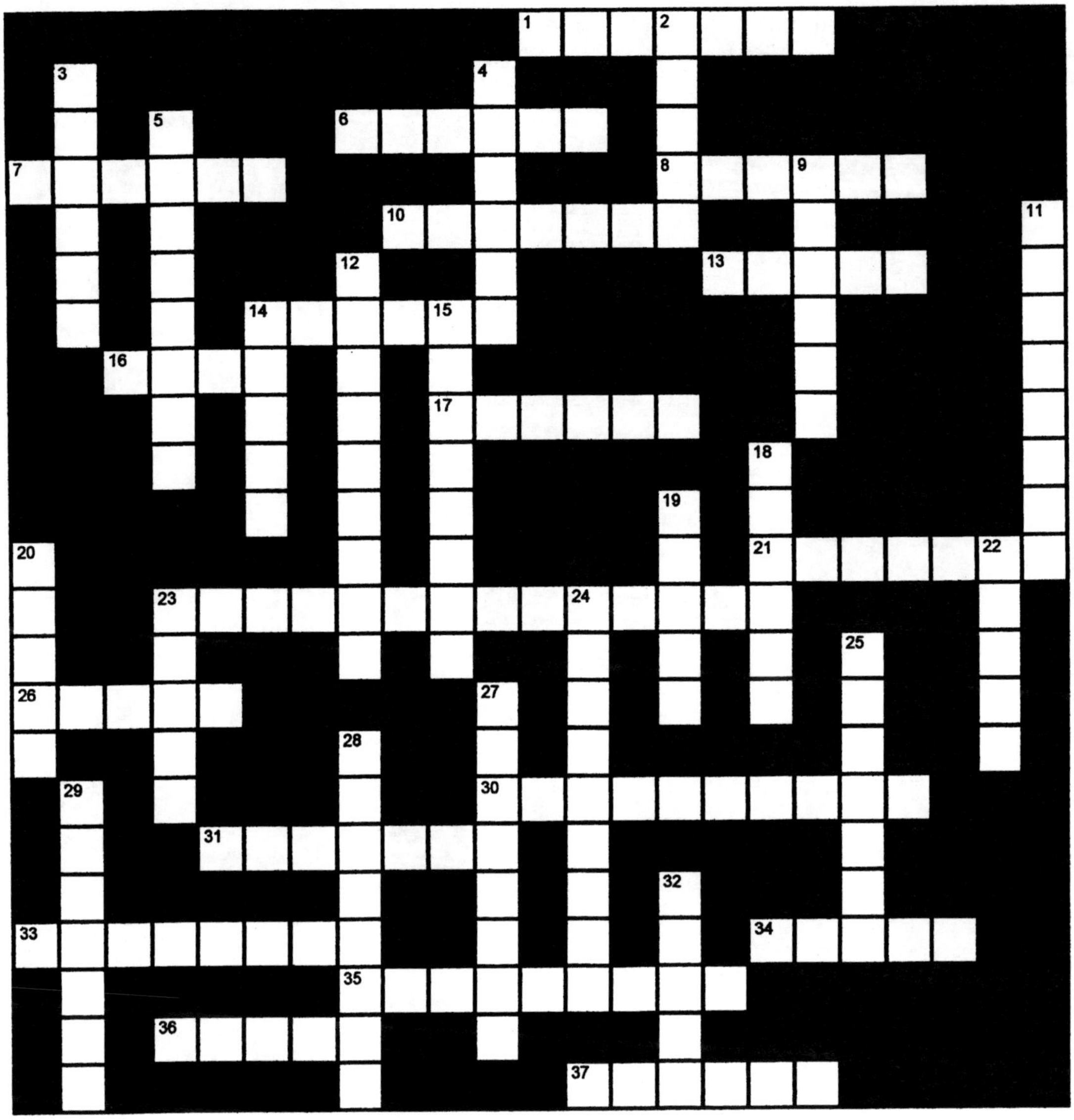

English Literature II Authors Puzzle Clues

Across

1.) He wrote *Pickwick Papers, Bleak House*, and *Great Expectations* and even traveled to America although American social reforms didn't live up to his expectations.

6.) Victorian poet and social critic who wrote "The Buried Life" and "Dover Beach."

7.) He shortened his name from the longer Polish version and wrote adventurous, searching novels like *Lord Jim* and *Heart of Darkness.*

8.) The surname of the 20th-century Welsh poet who wrote the famous villanelle "Do Not Go Gentle into that Good Night" about his father's death but later drank himself to death at a fairly young age.

10.) Wilkie ______ wrote 19th-century novels of mild supernaturalism such as *The Woman in White* and an early detective story such as *The Moonstone.*

13.) This poet, whose real name was George Gordon, became famous for poems with melancholy, contemplative, lonely heroes, but he was also capable of witty works like "Don Juan." He was a favorite of popular audiences and is credited with the observation, "I woke up one morning to find myself famous."

14.) The last name of several novel writing sisters who produced such works as *Jane Eyre* and *Wuthering Heights.*

16.) Victorian literary and social writer who tried to free women from tyrannical domestic subordination in *The Subjection of Women* (1869) and tried to free citizens from the pressure to conform to the ill-guided majority in *On Liberty* (1859).

17.) This Oxford professor felt like Anglicanism had become too watered down and secular, so he spearheaded a movement known as The Oxford Movement and also as Tractarianism in which he converted to Roman Catholicism and eventually was named a Cardinal.

21.) Romantic poet who wrote "Adonais" and "Ode to the West Wind."

22.) This woman writing in the late 1700's wrote a key, early treatise of feminism with *A Vindication of the Rights of Women.*

26.) Author of the still-popular Sherlock Holmes stories.

30.) The Romantic poet who lived the longest, publishing his final, large poem in 1850, the year of his death.

31.) 20th-century novelist who wrote *Howard's End, A Passage to India*, and a shorter book called *Aspects of the Novel.*

33.) Prolific Victorian novelist who wrote *The Warden, Barchester Towers, and The Last Chronicle of Barset.*

34.) 20th-century Irishman who wrote *Ulysses, A Portrait of the Artist as a Young Man*, and *Finnegans Wake.*

35.) Author of the Victorian novel *Vanity Fair.*

36.) Scottish author of Romantic novels such as *Ivanhoe and The Lady of the Lake*. Mark Twain felt that this man's works were so unrealistic that Twain named a sinking riverboat after him in *Huckleberry Finn.*

37.) Victorian writer who wrote about society and art in *Modern Painters* and *The Stones of Venice.*

Down

2.) Young writer of odes who tragically died prematurely of tuberculosis at only age 25 after writing about sadness and lyricism in "Ode on a Nightingale" and about the fleeting nature of time in "Ode on a Grecian Urn."

3.) Arnold wrote his elegy "Thyrsis" about the death of this young poet friend.

4.) 20th-century female writer who penned *To the Lighthouse, Mrs. Dalloway*, and a well known essay, "Shakespeare's Sister."

5.) Victorian poet known for great dramatic monologues like "The Bishop Orders his Tomb in St. Praxed's Church" and "Fra Lippo Lippi."

9.) Late 19th-century literary figure known for founding The Kelmscott Press and for creating fabric and art designs with oak leaves and repeating geometric shapes for the Arts and Crafts Movement.

11.) He wrote *Confessions of an English Opium-Eater* about the pains of opium addiction and attempts to kick the habit.

12.) 20th-century comic writer who wrote lyrics for early Broadway musicals. wrote about 90 books in his 93 years of life, was knighted in 1975, and created the successful comic literature duo of Jeeves and Wooster.

14.) He was an early Romantic writer and artist who wrote *Songs of Innocence and of Experience*.

15.) The most popular Victorian poet at the time of that era, he wrote the famous line "'Tis better to have loved and lost than never to have loved at all."00

18.) The author of such novels as *Pride and Prejudice* and *Emma*, this writer has garnered more acclaim in our lifetime than she did in her own.

19.) The real last name of novelist George Eliot.

20.) After writing powerful but dark tales such as *Tess of the D'Urbervilles* and *Jude the Obscure*, he quit writing novels and turned to poetry, writing such poems as "The Darkling Thrush."

22.) She wrote *Middlemarch.*

23.) The witty novelist and playwright from the late 19th century who wrote *The Importance of Being Earnest.*

24.) Although he indulged too much in the drug laudanum, he was nevertheless a powerful Romantic poet who wrote a supernatural tale about a sailor who is cursed because he took the life of a seabird for no good reason.

25.) Friend of 24 down who helped him locate a wife although the marriage was often unhappy later. The 2 men planned on moving to Pennsylvania together with their wives to start a utopian society there that they called a pantisocracy.

27.) This 20th-century novelist wrote *Sons and Lovers* and *The Rainbow.*

28.) This is the last name of a brother and sister from a literary and artistic family; a minor Victorian poet, she wrote "Goblin Market" while he wrote poetry and painted and helped found the artistic circle called The Pre-Raphaelite Brotherhood.

29.) The maiden name of Robert Browning's frail, poetic wife who was more popular as a poet during their lifetime than he was.

32.) Irish turn-of-the-century poet who wrote "Sailing to Byzantium" and "The Second Coming."